SpringerBriefs in Molecular Science

More information about this series at http://www.springer.com/series/8898

Eugene V. Babaev

Incorporation of Heterocycles into Combinatorial Chemistry

 Springer

Eugene V. Babaev
Chemistry Department
Moscow State University
Moscow
Russia

ISSN 2191-5407   ISSN 2191-5415   (electronic)
SpringerBriefs in Molecular Science
ISBN 978-3-319-50013-3   ISBN 978-3-319-50015-7   (eBook)
DOI 10.1007/978-3-319-50015-7

Library of Congress Control Number: 2016959806

Printed on acid-free paper

This Springer imprint is published by Springer Nature
The registered company is Springer International Publishing AG
The registered company address is: Gewerbestrasse 11, 6330 Cham, Switzerland

Contents

Introduction

Over the last decade, the phrase "combinatorial chemistry" has been often used in the lexicon of most foreign pharmaceutical companies with many chemists showing increasing interest in this topic. The term "combinatorial chemistry" has little to do with combinatorics (i.e., manipulation with abstract mathematical objects). It is interpreted as the science of combining diverse source of reagents to obtain diverse arrays of products. Such combination is not only (and not so much) virtual one. It is achieved through the practical implementation of tens, hundreds, and sometimes thousands of parallel chemical transformations with the formation of a vast number of end-products, usually called "libraries" of compounds.

A pure academic science synthesis of huge arrays of similar substances may be pointless, from a viewing point.[1] Meanwhile, this is the challenge with which all the urgency came to the fore in the late twentieth century. Mass objective of the synthesis was set by the practice of industry (chemical, especially pharmaceutical), where it is required to obtain materials or drug with the desired properties. The modern state of science still does not allow one to accurately predict the presence of useful properties based only on the structure of matter (or the composite). A more pragmatic and cost-effective is an empirical enumeration of properties in a large series that must somehow be synthesized.

The needs of the industry in this area have brought to the fore a number of issues, which appeared rarely in classical organic chemistry, but are key to combinatorial chemistry. First of all, these are rapid syntheses of many substances, usually complex in structure and relatively clean. Therefore, the tasks of combinatorial chemistry primarily involve the development of new efficient technologies for parallel synthesis and purification of substances. They also require modification of the usual chemical laboratory equipment to save time and space during the

---

[1]Suffice it to recall the skeptical expression "stupid synthesis of homologues" of Prof. A.N. Kost from Moscow University.

standard procedures of synthesis. Figuratively speaking, the main objective of combinatorial chemistry is to "compress space and time" during the synthesis of many compounds. Increasingly for this purpose are applied programmable robots that automate routine and repetitive procedures for loading, separation, and purification of many compounds simultaneously.

Thus, combinatorial chemistry is probably not a new scientific discipline, but a set of new ultraefficient technologies. Since not every reaction (easily occurring in a conventional bulb) can be effectively adapted to parallel synthesis of hundreds of vessels, the main scientific problem of combinatorial chemistry is the careful selection of the types of reagents and reactions, suitable for creating libraries. Particularly preferred are those methods where it is possible to shift the equilibrium toward the end-products without compromising their purity. This purpose is served by the methodology of using the "scavengers," solid and liquid perfluorinated phases, as well as the synthesis on a solid-phase support.

The main aim of the proposed book was therefore to fill the existing gap in the literature on modern technologies of combinatorial chemistry. The publications are designed to cover major inflorescence of modern combinatorial chemistry. A number of published articles are devoted to the "ABC" of parallel synthesis of small molecules (on solid and liquid phases). They are built on the principle of training "manuals" for students (with specific methodics and explanations), which allows their use in the workshops of the universities. The basis for these articles is the many years' experience of teaching students at the Department of Chemistry, Moscow State University, as well as cooperation between teachers of different universities in Russia and abroad.

The materials of the book will be useful both to specialists in the field of combinatorial chemistry and to new researchers, as well as clarify those who still have problems with the combination of the words "combinatorial chemistry".

The book is based on the author's experience of teaching combinatorial chemistry in the past twelve years. In Chap. 1, the solid-phase synthesis of unnatural heterocyclic alpha-amino acids is illustrated by practical examples starting from ABC of peptide synthesis. In Chap. 2, the technique of solid-phase synthesis is shown on various devices: billboards, tea bags, and lanterns. The major part here is demonstrated on the examples from the chemistry of heterocycles. In Chap. 3, the tools for accelerating chemical synthesis in liquid phase are reviewed. The technique of parallel refluxing (including microwave and flow technique) and parallel separation (filtration, centrifugation, evaporation, and chromatography) is described here. Chapters 4 and 5 illustrate the liquid-phase synthesis of heterocyclic scaffolds (reductive amination and Ugi reaction) with the use of semiautomated protocols. Finally, in Chap. 6, some personal impressions will be reviewed on the application of combinatorial heterocyclic chemistry in high school, including the concepts and the library design.

We greatly acknowledge the use of data published by Pleiades Publishing Ltd. [1–6].

References

1. Babaev EV, Ermolat'ev DS (2010) Basic techniques of working on a solid phase: from ABC of the peptide synthesis to libraries of non-natural amino acids. Russ J Gen Chem 80 (12):2572–2589
2. Babaev EV (2010) Solid-phase synthesis for beginners: choice of tools and techniques for implementation of multistage transformations. Russ J Gen Chem 80(12):2590–2606
3. Babaev EV (2010) Little artifices of huge libraries: secrets of parallel liquid-phase synthesis. Russ J Gen Chem 80(12):2607–2616
4. Ivanova NV, Tkach NV, Belykh EN, Dlinnykh IV, Babaev EV (2010) Reductive amination with a scavenger: the most "Combinatorial" of two-component reactions. Russ J Gen Chem 80(12):2617–2627
5. Mironov MA, Babaev EV (2010) A parallel Ugi reaction at students laboratories in the ural and Moscow. Russ J Gen Chem 80(12):2647–2654
6. Babaev EV (2010) Combinatorial chemistry in higher school: ten-year experience of research, educational, and managerial projects. Russ J Gen Chem 80(12):2655–2670

Chapter 1
Peptide Synthesis of Libraries of Non-natural Amino Acids

The present chapter has a slightly unusual structure. It is based on the experiments aimed at developing a training task for special students' laboratory course in combinatorial chemistry in the Moscow State University (MSU). From one side, it combines features of a review (carefully selected publications on the techniques of peptide synthesis). From the other side, these are experimental protocols tested step by step by the author. Finally, it is an original synthesis of a previously inaccessible family of non-natural amino acids with a well-defined structural motif of medicines. In its chemical essence, the chosen reaction sequence is the following classical scheme of peptide synthesis: (1) immobilization of an *N*-substituted amino acid on a solid support; (2) removal of the protective group; (3) modification of the amino acid NH_2 group; and (4) removal of the modified amino acid from the support. The original element of this task and its peculiar zest is the stage in which the amino acid is modified by hetarylation—a reaction fairly rarely used for the preparation of non-natural amino acids [1, 2].

First, we substantiate why the molecular structures in the obtained library may suggest biological activity. Then, we analyze features of each of the four stages in terms of methodology and published techniques. In conclusion, we give concrete or practical recommendation and provide experimental protocols, as well as evidence for the structure and purity of the synthesized compounds. The principal methodical feature of the presented protocols consists of those which require no special equipment and can be reproduced in usual students' laboratory works.

1.1 Statement of the Problem

At present, the biological activity of 3-guanidinopropionic acid attracts the interest of medical chemists due to its expressed antidiabetic activity and, simultaneously, metabolic instability in human's body. Among drugs applied, the most known is tiformin (**A**) which is 3-guanidinopropionamide. Larsen and coworkers [3, 4],

© The Author(s) 2017   1
E.V. Babaev, *Incorporation of Heterocycles into Combinatorial Chemistry*,
SpringerBriefs in Molecular Science, DOI 10.1007/978-3-319-50015-7_1

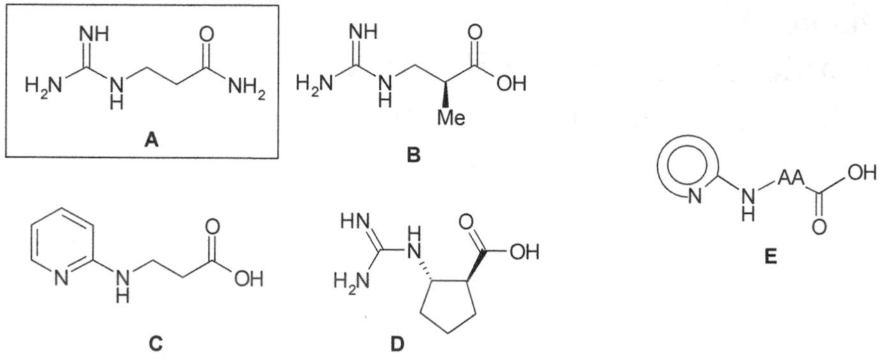

Scheme 1.1 Tiformin (A) and its analogs (B-E)

optimizing the structure of 3-guanidinopropionic acid and searching for its biologically active analogs, synthesized and tested for biological activity a series of
homologs and bioisosteres of this acid (Scheme 1.1):

It was shown that activity is largely associated with the amidine fragment of
3-guanidinopropionic acid. On the other hand, point mutations and substitution of
the guanidine fragment resulted in partial or complete loss of activity. By contrast,
the authors noted that α-alkylation (compound **B**), ring closure in the amidine or
guanidine fragment (compound **C**), and enhancement of the conformational rigidity
of the alkyl chain (compound **D**) sometimes enhanced the antihyperglycemic
activity. It follows from these results that one of the approaches to structural
modification of 3-guanidinopropionic acid might involve a systematic screening of
heterocycles for replacing the guanidine fragment. The bioisosteric replacement of
the carboxy group with wide variation in the structure of the alkyl chain leads to
similar effect. Thus, we can set the task to develop a method of synthesis of a wide
series of heterocyclic amino acids of the general formula **E**, having the guanidine
and amidine fragments incorporated in an α-aminoheterocycle and an amino acid
residue (AA). This class of compounds is well documented, but even though the
problem of preparation of *N*-hetarylamino acids is quite urgent, no universal synthetic approach to them is still available. The literature review showed that the
principal method of synthesis of such *N*-hetarylamino acids is the *N*-alkylation of
α-aminoheterocycles and nucleophilic substitution in the heterocyclic nucleus with
amino acid derivatives.

In comparing these two methods, one should consider the facts that amino acids
and their protected derivatives are commercially available. Subsequently, the second method is a more expedient synthetic approach to a wide range of hetarylamino
acids. This approach makes use of amino acid derivatives and involves aromatic
nucleophilic substitution in hetero-rings containing a readily leaving group in the
α-position. It is well known that free amino acids are difficult to apply, since they

tend to enter side reactions. Therefore, amino acids should be preliminarily protected. Their intermolecular interaction should be minimized.

A radical approach to the problem of synthesis of a series of derivatives **E** by a
unified strategy might be provided by solid-phase synthesis. Apparently, the
solid-phase synthetic approach allows to block the carboxylic function of the amino
acid by forming an ester or amide bond with a polymer support on the one hand.
Through this process, the intermolecular interactions of amino acids are excluded
by forming diketopiperazines and oligopeptides. On the other hand, by using a
polymer support, one can prepare free acids and esters, as well as amides. In this
manner, the potential of this method is extended.

As a result, we consider the most acceptable approach based on the aromatic
nucleophilic substitution in heterocycles under the action of amino acids immobilized on a polymer support. (This reaction can also be treated as N-hetarylation of
amino acids.) As the heterocycles for hetarylation, we chose 2-halogenated ones:
pyrimidines, 5-nitropyridine, and 5-nitrothiazole. All these halo derivatives fairly
readily react with primary and secondary aliphatic amines, and the results are
thoroughly examined and published. Thus, the chosen strategy for the synthesis of
N-hetarylamino acids can be represented by the following Scheme 1.2 (here and
hereinafter, AA and PG stand for amino acid residue and protective group,
respectively).

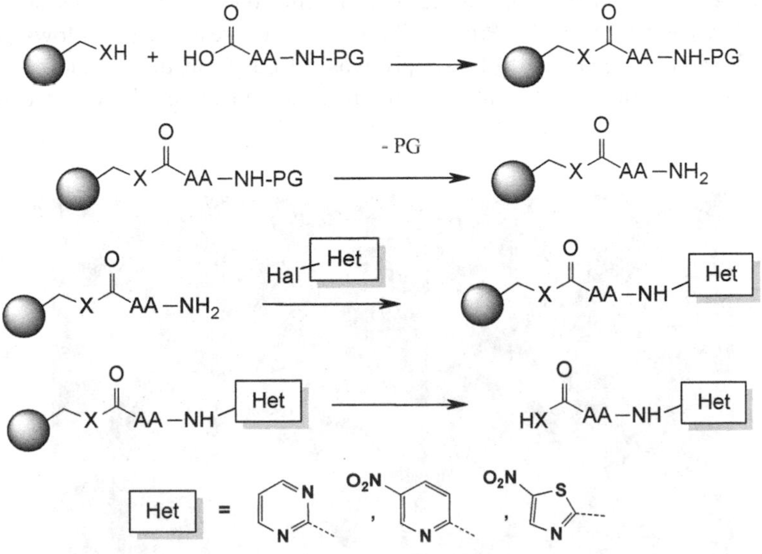

Scheme 1.2 Chosen strategy for the synthesis of N-hetarylamino acids

1.2 Solid-Phase Organic Synthesis

The methodology of modern solid-phase organic synthesis (SPOS) is in many respects adopted from the solid-phase synthesis of peptides. However, there is scarce information in the literature on the specific features of solid-phase organic synthesis and the advantages it offers over the traditional liquid-phase synthesis. Therefore, we consider it appropriate to give a brief characteristic of solid-phase synthesis and polymer supports used in this method [5].

1.2.1 Specific of SPOS

The aim of SPOS is to accelerate the synthetic process due to facilitated isolation and exclusion of additional purification of both intermediate and target products. The term *solid phase* relates in essence to physical characteristics of a substance on a support, since chemical reaction on a polymer support occurs in a single phase, specifically in a solution. In an appropriate solvent, a polymer swells (Fig. 1.1) and converts into a low viscous but strongly structured gel (cross-linked polymers) or dissolves (non-cross-linked polymers). Therefore, the synthesis occurs at an ultra-micro-heterogeneous level, i.e., in an almost homogeneous system.

SPOS requires the use of a polymer support (resin **S**) and its attached linker **L** (Scheme 1.3). The first stage (Stage 1) involves the attachment of a molecule of substrate **A** to the linker. Molecule **X** is immobilized but preserves its ability to react with reagent **Y** (Stage 2). Product **XY** stays on the resin, which allows excess reagent **Y** (and by-products) to be simply washed out from the product. (Further reagents can be added so that the structure of substrate **X** is successively

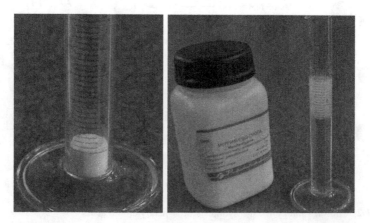

Fig. 1.1 Swelling of the Merrifield resin in CH_2Cl_2 (as seen, 1 ml of a granular material increases its volume 3 times)

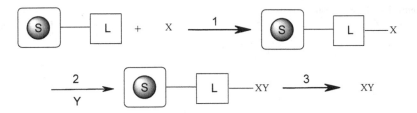

Scheme 1.3 Overview of SPOS (see text)

complicated; the key point is that the linker in these reactions remains untouched.)
A bifunctional linker **L** is selected so that it is bonded to resin **S** stronger than
substrate **X**. Then, at the last stage (Stage 3), the target compound **XY** can be
separated from the resin by breaking its bond with the linker. It is clear that the
L–XY bond should be destroyed in mild conditions, not damaging both the compound itself (bond **X–Y**) and the linker–resin contact (bond **L–S**).

Thus, in an ideal case, by washing the resin after each stage and breaking the
bond with the support, one can obtain a pure compound. It is natural to suggest that
using a large excess of reagents (with subsequent separation from the resin) can be
allowed in many cases to shift the chemical equilibrium to the target product and
shorten the synthesis timing.

The disadvantages of SPOS include (1) the necessity of using a fairly large
excess (2–30 equivalent) of reagent; (2) difficulties in the identification of intermediate products; and (3) a fairly high cost of modified polymer supports, which
depend on the cost of linkers.

1.2.2 Polymers Used

Chloromethylated polystyrene (cross-linked with a small amount of divinyl benzene), introduced in the practice of organic synthesis by Merrifield (so-called
Merrifield resin), is the most accessible of polymer supports (Scheme 1.4). The
chloromethyl group of the polymer allows easy alkylation of carboxylic acid
derivatives and their fixation on the resin. However, the resulting ester bond (of the
benzyl type) cleaves in fairly rigid conditions (conc. HF, HBr, alkali metal alcoholates, boiling with amines, etc.). In the peptide synthesis, the peptide bond with
the Merrifield resin can be cleaved with HF in thioanisole. Such conditions are
unfavorable for many compounds, and therefore, the Merrifield resin is modified to
facilitate the removal of products from the support. For example, the functional
group can be "elongated" by replacing the chlorobenzyl group with a p-benzy-
loxybenzyl (Wang linker) or benzhydryl group (Rink linker) in Scheme 1.4 [5].
Such benzyl esters are cleaved with acids in milder conditions. Additional introduction of the electron–donor methoxy groups in the benzene ring further enhances
the stability of the benzyl and benzhydryl cations formed by an acid hydrolysis of

Scheme 1.4 Typical resins used in SPOS

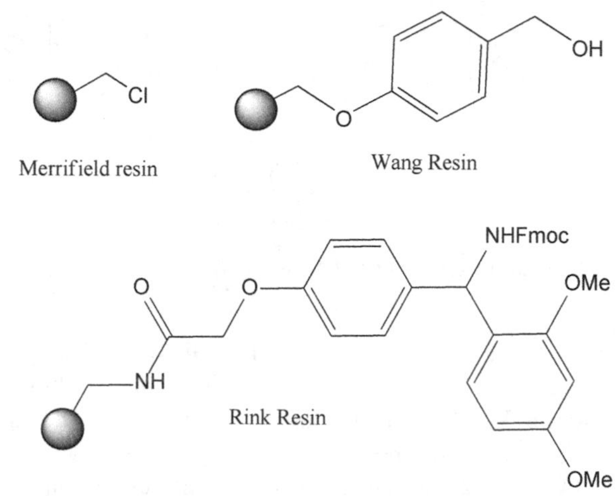

Merrifield resin

Wang Resin

Rink Resin

the linker. This allows the Wang linker to be cleaved with 50% CF_3COOH (TFA) and the Rink linker, with 20% TFA. Moreover, modification of the support extends the range of potential reaction products. Thus, with the Wang resin, one can obtain acids and alcohols and with the Rink resin, primary and secondary amides.

1.2.3   Methodology and Principal Stages of SPOS

The task we have set for ourselves required a polymer support with a grafted amino acid to be reacted with a heterocycle, activated to substitution. Let us consider in more detail the methodical issues of how to prepare amino acids immobilized on polymer supports.

1.2.3.1   Stage 1: Immobilization of *N*-Substituted Amino Acids on Polymer Supports

The first stage of our scheme involves immobilization of amino acids on polymer supports (reaction 1, Scheme 1.5).

To avoid such side, processes as oligopeptide formation amino acids are preliminarily protected. As a rule, *N*-substituted amino acids are used, and the resulting amino acid–support bond is an amide or ester bond. In SPOS, amino groups are most commonly protected by carbamate-type groups, like *tert*-butoxycarbonyl (Boc) [6] and 9*H*-fluorenylmethoxycarbonyl (Fmoc) [7] (X stands for a group to be protected) in Scheme 1.6.

It should be noted that the choice of protective group is determined by the type of the polymer support. Immobilization conditions for protected amino acids vary

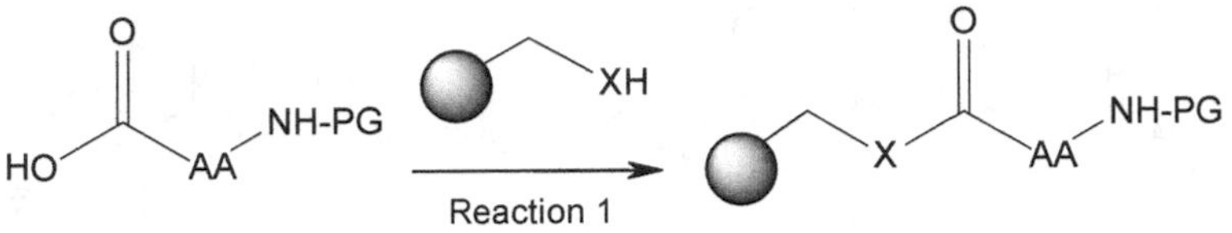

Scheme 1.5 First step: immobilization of aminoacid on resin. X = NH, O

Scheme 1.6 Carbamate-type groups: Boc and Fmoc

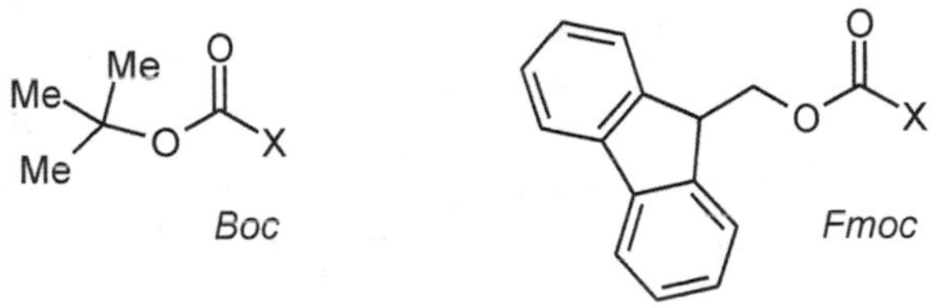

from one polymer support to another. Boc-amino acids are immobilized on the Merrifield resin (chloromethylated polystyrene) in situ as cesium salts, by adding a suspension of cesium carbonate in DMF and a potassium iodide catalyst (Scheme 1.7).

The reagent to the support ratio is chosen individually in each case and spans the range 1.5–4. Immobilization of Fmoc-amino acids on the Wang support (X = O) to form a benzyl ester linker is performed by the carbodiimide method, by treatment with diisopropylcarbodiimide (DIC) in the presence of a 4-(dimethylamino)pyridine (DMAP) catalyst (Scheme 1.8, left). The immobilization reaction with sterically uncongested amino acids occurs at a room temperature. Sterically congested amino acids are immobilized at 40–60 °C for 2 days (two immobilization cycles).

The immobilization of Fmoc-amino acids on the Rink resin (X = NH) to form a benzhydryl amide linker is performed in the presence of the Castro's reagent {(1*H*-1,2,3-benzotriazole-1-yloxy)tris-(dimethylamino)-phosphonium (BOP)}, diiso-propylethylamine (DIEA) as a base, and 1-hydroxy-benzotriazole (HOBt) as a catalyst (Scheme 1.8, right). The reaction occurs at room temperature for 2 h for sterically uncongested (or 4–6 h with sterically congested) amino acids.

1.2.3.2 Stage 2: Deblocking Substituted Amino Acids on Polymer Supports

The second stage of our synthetic scheme (after immobilization of a protected amino acid) involves the removal of the protective group to activate the amino group (reaction 2, Scheme 1.9).

The Boc and Fmoc protections are removed in different ways. The removal of the Boc protection in amino acids on the Merrifield resin is performed by treatment with 50% trifluoroacetic acid in dichloromethane for 30 min (Scheme 1.10).

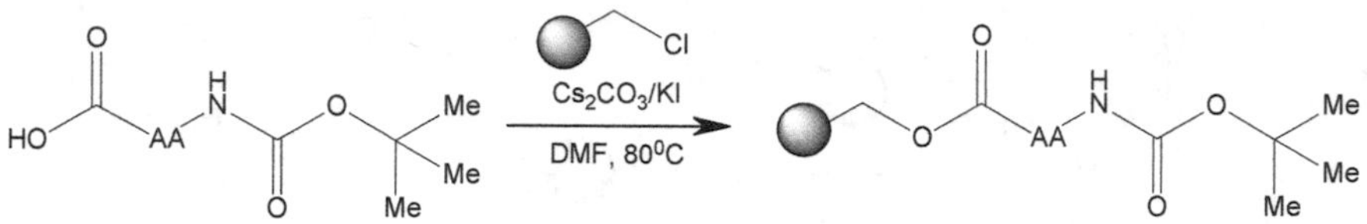

Scheme 1.7 Immobilization of Boc-amino acids on the Merrifield resin

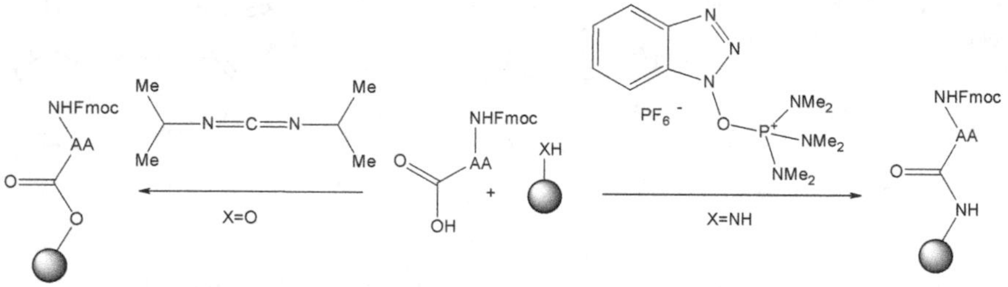

Scheme 1.8 Immobilization of Fmoc-amino acids on the Wang (*left*) and Rink (*right*) resins

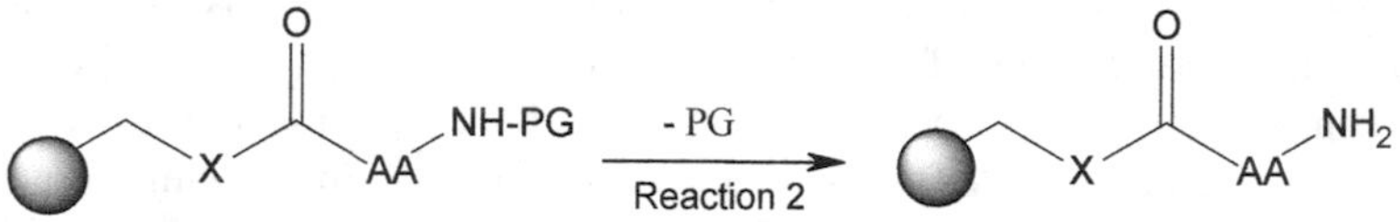

Scheme 1.9 Second step: deblocking of the amino acid

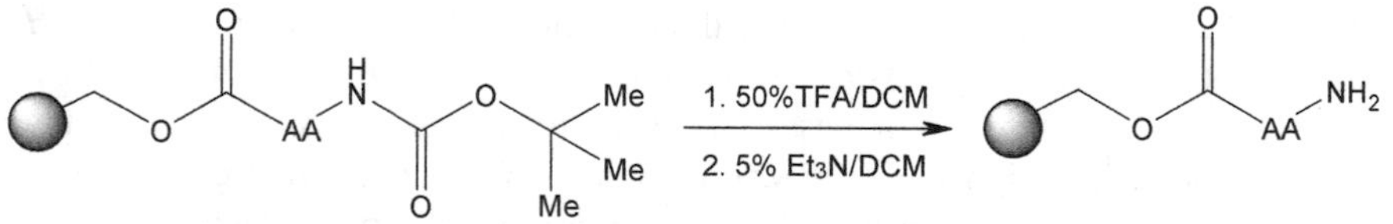

Scheme 1.10 Deblocking of Boc-protected amino acid

Under these conditions, the Merrifield linker remains intact. After deprotection, the resin is washed with a triethylamine solution to remove trifluoroacetic acid. The Fmoc protection in amino acids on the Wang (X = O) and Rink (X = NH) supports is removed with a 20% piperidine solution in DMF for 40–50 min, in Scheme 1.11. A considerable weight loss of the resin after removal of the Fmoc protection can form a basis for the gravimetric assessment of the degree of immobilization of protected amino acids at the first stage of solid-phase synthesis.

Consecutive treatment of the resin with a solution of piperidine in dimethyl phthalate is recommended: first, for 5–10 min. and then, for 30 min, in a fresh

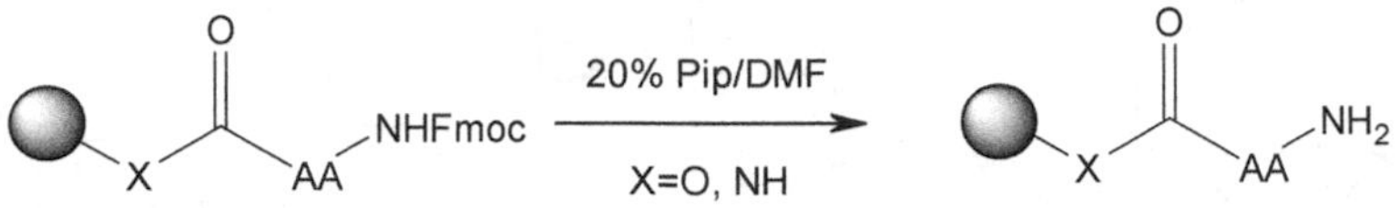

Scheme 1.11 Deblocking of Fmoc-protected amino acid

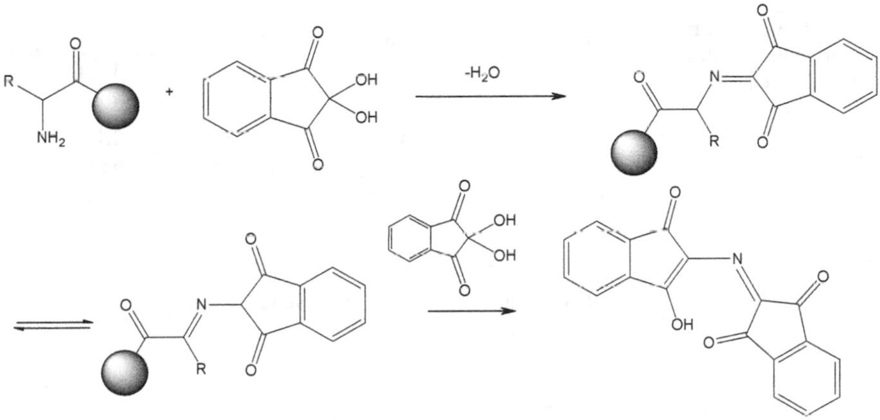

Scheme 1.12 Color ninhydrin reaction (Kaiser test) for the free amino group

solution. After deprotection, the resin is washed no less than 4 times with dimethyl phthalate to remove Fmoc decomposition products. The progress of acylation on supports or the removal of protective functions from amino acids can be followed by means of the Kaiser test.

Kaiser Test for Amino Group. Analysis of resins after a reaction which results in disappearance or appearance of a free amino group is readily accomplished by means of the ninhydrin (Kaiser) test [8]. This test can be both qualitative and quantitative, and it is a fairly sensitive color reaction for amino group (Scheme 1.12). A deep dark blue color of the resin and solution provides clear evidence for the presence of a primary amine function on the resin. If the color remains yellow, one can conclude that such function is absent. With secondary (proline) or sterically congested (phenylalanine and β-phenylalanine) amines, the resin and solution acquire a dark red color, which is typical in such cases.

1.2.3.3 Stage 3: Nucleophilic Substitution in Heterocycles, Involving Polymer-Immobilized Amino Acids

The next stage in our planned protocol involves aromatic nucleophilic substitution. The nucleophile is the grafted amino acid, and the activated heterocycle is present in solution (reaction 3, Scheme 1.13).

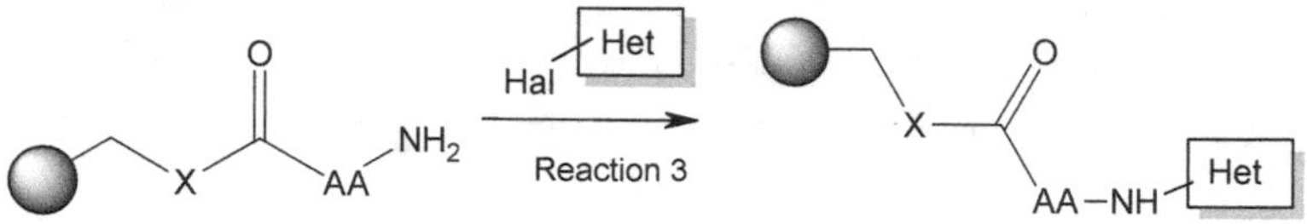

Scheme 1.13 Third step: arylation of the NH_2 group of amino acid by active heterocycle

Table 1.1 Reaction rates of halopyrimidines with piperidine in ethanol

Pyrimidine	Relative reaction rates [9]		
	20 °C	30 °C	40 °C
2-Br	1	2.05	4.1
2-Cl	0.49	1.03	2.11
2-F	66.23	117.7	236
2-I	0.3	0.64	1.35

Table 1.2 Relative nucleophilic substitution rate constants in activated pyrimidines under the action of cyclohexylamine (29 °C, EtOH)

Pyrimidine	Relative rate constant [10]
2-Cl	1.0
2-MeSO	2.2
2-MeSO$_2$	4.7
4-MeSO	1.1×103
4-MeSO$_2$	6.7×103
2-PhSO$_2$	4.5

Most nucleophilic substitutions on supports are accomplished similarly to liquid-phase reactions. However, the reaction temperature should not be above 120 °C, since at higher temperatures the polystyrene base of the carrier starts to degrade. Conditions of supported reactions should also preserve the linker. In selecting appropriate activated heterocyclic substrates, one should take into account the nature of the leaving group in the heterocycle. Tables 1.1 and 1.2 list the activities of pyrimidines with different leaving groups in the 2 and 4 positions. As shown in Tables 1.1 and 1.2, the rate of substitution in 2-halopyrimidines is the highest for the fluorine derivative, the 4th position is more active than the 2nd, and sulfonyl groups are preferred over chlorine. The listed data are helpful in selecting other hetarylating substrates.

1.2.3.4 Stage 4: Removal of Target Compounds from Polymer Supports

Most linkers in solid-phase organic synthesis are cleaved in an acid medium (re-action 4, Scheme 1.14) [11].

The resistance of linkers to acids is sharply decreased in going from the Merrifield to Wang and Rink resins. The Rink linker is cleaved in milder conditions

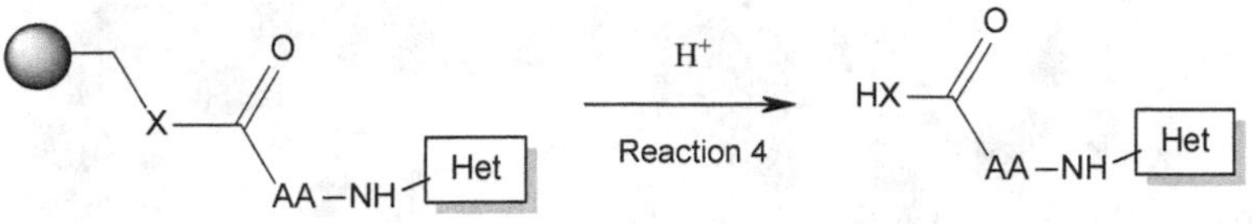

Scheme 1.14 Fourth stage: removal of the final compound from the resin

Scheme 1.15 Use of
MeONa during the removal
produced and ester

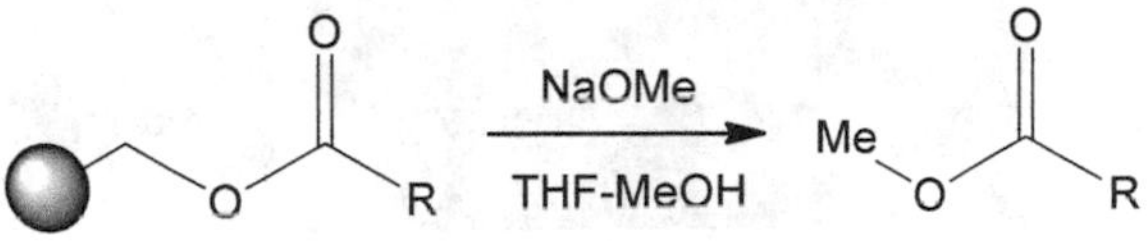

(10–20% CF$_3$COOH) than the Wang linker (50% CF$_3$COOH). The Merrifield resin
is passive in such conditions, and it is cleaved by transesterification forming an acid
ester [12] in Scheme 1.15. We would like to reiterate that the nature of the linker
determines the type of the terminal function in the molecule removed from the
support. The Wang resin allows the production of acids, and the Rink resin allows
the production of amides.

1.2.3.5   Equipment and Materials

Experimental implementation of the above-described four-stage sequence requires
no specific laboratory equipment. Reactions requiring no heating are convenient to
perform in capped vials or plastic syringes with a porous partition (Fig. 1.2a),
minimizing manipulations at the stages and filtering and resin washing. If heating is
required, glass vials (desirably, from a heat-resistant glass and hermetic screw caps)
can be fixed in a temperature-controlled shaker (we used a low-cost domestic
device shown in Fig. 1.2b).

The set of reagents includes three basic components: a "school" set of amino
acids; Merrifield and/or Rink resin; and simple substituted heterocycles (vide infra).
The commercial immobilized amino acids can be used. In our case (with account
for protection/immobilization of amino acids); such reagents as di-*tert*-butyl
dicarbonate and 9-fluorenylmethyl chloroformate (introduction of Boc and Fmoc
protections) and reagents for immobilization (Cs$_2$CO$_3$/KI, HOBt, or BOP) are
needed. Ninhydrin is required for the Kaiser test and CF$_3$COOH and piperidine—
for the removal of protective groups. At certain stages, we took DIEA. A large
consumption of solvents, specifically DMF and dichloromethane (DCM), for
washing resins should also be taken into account.

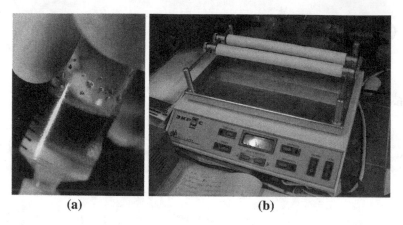

Fig. 1.2 Equipment for SPOS: **a** plastic syringe with a porous partition and **b** temperature-controlled shaker

1.3 Practical Implementation of the Task

1.3.1 Stage 1: Immobilization of Amino Acids on Supports

The starting materials for Stage 1 are N-protected amino acids. Boc- and Fmoc-amino acids are commercially available, but fairly expensive. By this reason, we considered it reasonable to prepare Boc and Fmoc derivatives of the selected amino acids.

Preparation of Boc-Amino Acids **I**. Amino acids are heated with di-*tert*-butyl dicarbonate in aqueous dioxane in the presence of NaOH [13] in Scheme 1.16. Amino acid, 0.04 mol, was added to a solution of 1.6 g of NaOH (0.04 mol) in 40 ml of water. The solution was shaken for 10 min, cooled to 10 °C, and a solution of 6.55 g (0.03 mol) di-*tert*-butyl dicarbonate in 30 ml of dioxane was added over the course of 15 min. After 20 min, the solution was heated with caution to 40–60 °C (CO_2 evolution is initiated) and stirred for 16–40 h until di-*tert*-butyl dicarbonate was no longer detected by TLC. The reaction mixture was poured into 100 ml of ice water, and the pH of the solution was brought to 3 by adding 10% HCl in small portions. After chloroform extraction (2 × 100 ml), the organic phases were combined, and the solution was removed on a rotary evaporator. The remaining colorless oil was crystallized by trituration in hexane in an ultrasonic bath. The products were dried in a vacuum oven and purified, if required, by recrystallization from hexane/ethyl acetate. TLC control: chloroform–methanol–acetic acid, 9:1:0.1. In this way, we obtained compounds **Ia–Ih**: Boc-glycine, Boc-L-alanine,   Boc-β-alanine,   Boc-L-phenylalanine,   Boc-β-phenylalanine, Boc-L-proline, Boc-5-aminovaleric acid and Boc-4-(aminomethyl) benzoic acid. The purity of the products was assessed by NMR spectroscopy.

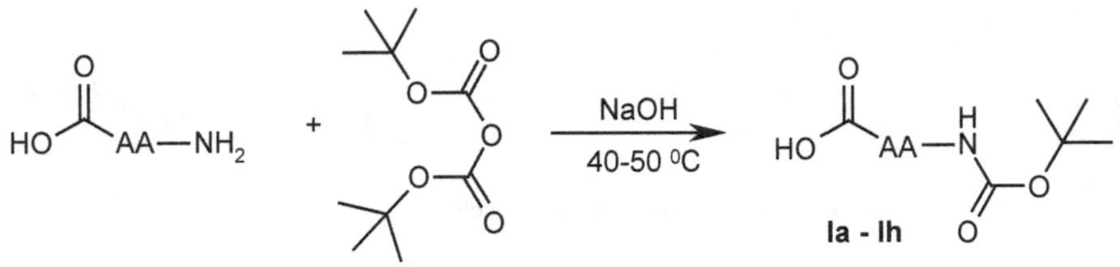

Scheme 1.16 Preparation of Boc-amino acids

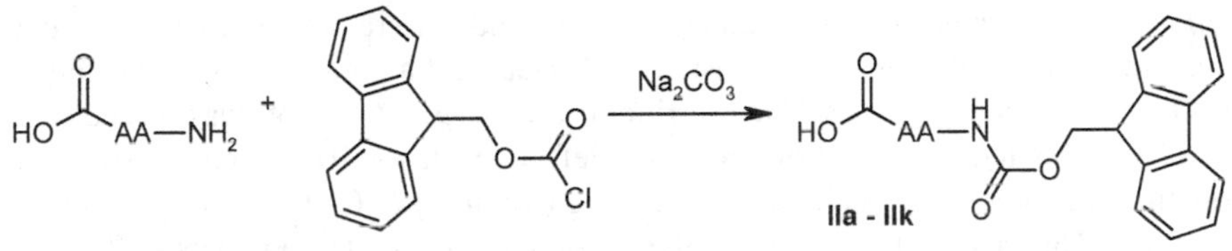

Scheme 1.17 Preparation of Fmoc-amino acids

Preparation of Fmoc-Amino Acids **II**. The Fmoc protection was introduced using 9-fluorenylmethyl chloroformate [7] in the presence of sodium carbonate in Scheme 1.17. By replacing sodium carbonate with DIEA [14], we could decrease the yield of 9-fluorenylmethyl chloroformate hydrolysis products (dibenzofulvene and its polymerization products, as well as 9-fluorenylmethanol).

Method A. Amino acid (20 mmol) was dissolved (or partially suspended), with stirring and cooling on an ice bath, in 50 ml of 10% Na_2CO_3. After that, dioxane, 20 ml, was added and then, slowly, a solution of 5.7 g (22 mmol) of 9-fluorenylmethyl chloroformate in 50 ml of dioxane. The mixture was shaken for 1 h at 0 °C and then at a room temperature from 3 to 8 h, depending on the amino acid. The reaction mixture was diluted to 1 l with ice water and extracted with ether (3 × 200 ml) to remove chloroformate decomposition products: dibenzofulvene and its polymers (R_f 0.85–0.9; ethyl acetate/hexane, 1:1). The aqueous layer was cooled on an ice bath, acidified with conc. HCl to pH 2.0, and extracted with ethyl acetate (4 × 200 ml). The combined organic phases were washed with 0.1 N HCl and water, dried over Na_2SO_4, and evaporated on a rotary evaporator to obtain a viscous colorless oil which could be crystallized in *n*-heptane on an ultrasound bath. The resulting Fmoc-amino acids were finally purified by recrystallization from ethyl acetate/hexane. TLC control: chloroform/ethanol, 3:1 (system I), and chloroform/methanol/acetic acid, 9:1:0.1 (system II).

Method B. Here, we used DIEA as a base and at least a 20% excess of the base and amino acid with respect to chloroformate, which allowed us to increase the total reaction yield and significantly decrease the formation of by-products.

Thus, we could prepare, in fairly high yields, the Fmoc derivatives of glycine, β-alanine, 4-aminobutyric acid, 5-aminovaleric acid, L-alanine, L-phenylalanine, β-phenylalanine, L-asparagine, DL-valine, L-proline, and 4-(aminomethyl) benzoic

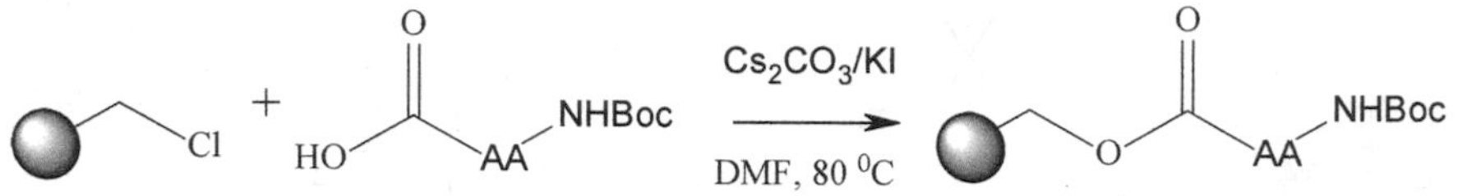

Scheme 1.18 Immobilization of Boc-protected amino acid

acid. The resulting products **II** were characterized, and their purity was assessed by NMR spectroscopy.

Immobilization of Boc- and Fmoc-protected amino acids was performed by standard procedures. Boc-substituted amino acids were immobilized on the Merrifield acid in the presence of cesium carbonate and a potassium iodide catalyst at 80–85 °C in Scheme 1.18. The Merrifield resin (1 g, 1.6 mmol) was placed in a 30-ml vial, poured with 3 ml of DMF, and left to swell for 10–15 min. A solution containing 5 mmol of a Boc-amino acid, 1.6 g (5 mmol) of Cs_2CO_3, and a catalytic amount of NaI (1 mmol, 0.17 g) was then added to the swollen polymer so that the total volume of the reaction mixture was no more than 10 ml. Tightly capped vials were heated in an oil bath at 80–85 °C for 12 h with intermittent shaking. The resin was then separated on a glass frit, washed successively with DMF (20 ml), 10% NH_4Cl (3 × 30 ml), DMF (20 ml), DCM (2 × 30 ml), and MeOH (2 × 15 ml), dried in a vacuum oven and weighed.

The yields (%) at the stage of immobilization of protected amino acids on the Merrifield resin were as follows:

Glycine (Gly)	94
β-Alanine (β-Ala)	95
5-Aminovaleric (5-Ava)	92
L-Alanine (L-Ala)	96
L-Phenylalanine (L-Phe)	82
DL-β-Phenylalanine (DL-β-Phe)	76
L-Proline (L-Pro)	95
4-(Aminomethyl)benzoic (Amb)	97

The Fmoc-substituted amino acids were immobilized on the Rink resin by treatment with DIC in the presence of 1-hydroxybenzotriazole in Scheme 1.19:

Carbodiimide Method. A dry Rink–NH_2 resin (1 g) was placed in a 30-ml vial, poured with 3 ml of a solution of 0.46 g (3 mmol) of 1-hydroxy-1*H*-1,2,3-benzotriazole in 1:1 DMF/DCM, and left to swell for 10 min. After that 3–4 of a 1:1 ml DMF/DCM solution containing 3 mmol of a *Fmoc*-amino acid and 3 mmol (0.38 g) of DIC is added to the resin. If a precipitate is formed, more DMF was added. Tightly capped vials were agitated on an orbital shaker for 15–30 h. The reaction progress was followed by the Kaiser test. As a rule, the reaction time was 4–5 h with unbranched amino acids and 6–9 h with sterically congested amino acids. After the

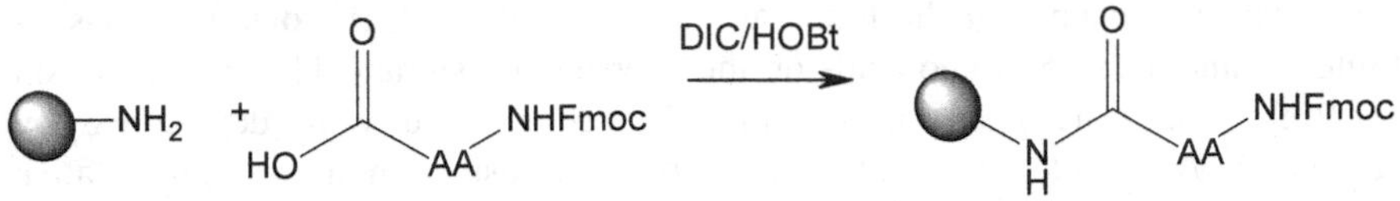

Scheme 1.19 Immobilization of Fmoc-protected amino acid

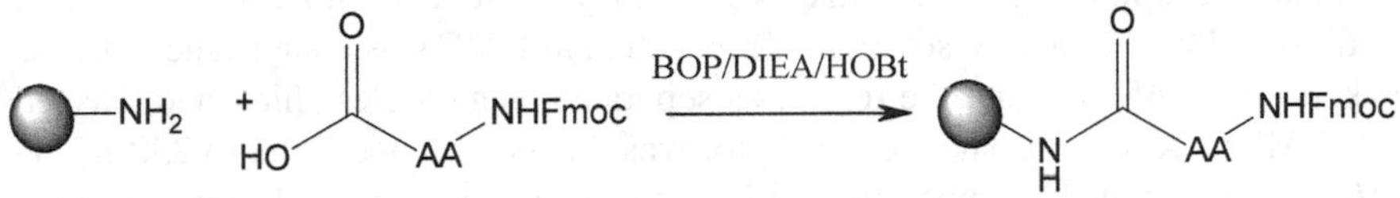

Scheme 1.20 Immobilization of Fmoc-protected amino acid using BOP/HOBt

reaction completion, the resin was washed in succession with DMF (3 × 20 ml), DCM (2 × 20 ml), and MeOH (2 × 10 ml) and dried in a vacuum oven. The resin weight is always nicely correlated with the calculated reaction yield.

Sterically congested amino acids, such as L-phenylalanine, L-asparagine, DL-valine, and β-phenylalanine, were immobilized on the Rink resin by means of (1*H*-1,2,3-benzotriazol-1-yloxy)tris (dimethylamino)-phosphonium hexafluorophosphate (BOP) and diisopropylethyl-amine (DIEA) in the presence of 1-hydroxybenzotriazole (HOBt), Scheme 1.20. A solution (6–7 ml) of 1.33 g (3 mmol) of BOP, 3 mmol of Fmoc-amino acid, and 0.78 g (6 mmol) of DIEA in 1:1 DMF/DCM was added to deblocked Rink resin (1 g). The mixture was agitated on an orbital shaker for 5 h. The resin was separated, washed with DMF (2 × 20 ml), DCM (2 × 30 ml), and MeOH (1 × 15 ml), and dried in a vacuum oven. The reaction progress was followed by the Kaiser test.

The yields at the stage of immobilization of protected amino acids on the Rink resin were as follows, %:

Glycine (Gly)	99
β-Alanine (β-Ala)	100
4-Aminobutyric (GABA)	98
5-Aminovaleric (5-Ava)	99
L-Alanine (L-Ala)	98
L-Phenylalanine (L-Phe)	89
DL-β-Phenylalanine (DL-β-Phe)	94
DL-Valine (DL-Val)	90
L-Proline (L-Pro)	98
L-Asparagine (L-Asp)	87
4-(Aminomethyl)benzoic (Amb)	95

In total, we obtained at the first stage 19 supports (1–4 g), containing N-substituted amino acids: 8 amino acids on the Merrifield resin and 11 amino acids on the Rink resin. The yields at the immobilization stage were determined by gravimetry. As a rule, the carbodiimide method results in almost quantitative immobilization yields on the Rink resin. With sterically congested amino acids, the yields are slightly lower.

Determination of the Immobilization Degree of Fmoc-Amino Acids. To a resin obtained by coupling 1 g of the Rink–NH$_2$ resin with Fmoc-amino acids in standard conditions, 14 ml of a 25% solution of piperidine in DMF was added. The vial was shaken for 40–45 min, and the resin was separated using a glass filter and washed with DMF (3 × 10 ml). The mother liquor was diluted with ice water to 250 ml. 1-(9H-Fluoren-9-ylmethyl)-piperidine which precipitated as white flakes was equally filtered, washed with water, and dried in a vacuum, mp 119–120 °C. ^{1}H NMR spectrum (DMSO d_6), δ, ppm: 7.77 m (H$_{16}$, H$_{20}$, 2H), 7.33 d (H$_{13}$, H$_{17}$, 2H), 7.15 m (H$_{14}$, H$_{15}$ H$_{18}$, H$_{19}$, 4H), 5.21 t (CH, $J = 5.5$ Hz, 1H), 3.71 d (CH$_2$, $J = 5.15$ Hz, 2H), 2.50 m (2CH$_2$, 4H), 1.68 m (3CH$_2$, 6H). The yield of the immobilization reaction of Fmoc-amino acids on the Rink resin was calculated by the following formula:

$$Yield = \frac{m(\text{precipitate}) \cdot 1000}{M.C} \cdot 100\%,$$

where m is the weight of the precipitate, $M = 263.4$, molecular weight of 1-(9H-fluoren-9-ylmethyl) piperidine, and $C = 0.73$, capacity of the Rink resin, mmol/g.

1.3.2   Stage 2: Removal of Protective Groups

The stage of immobilization is immediately followed by the stage of protection removal. Free amino groups on the resin are readily determined by the Kaiser test (vide supra).

Kaiser test. We prepared 3 vials (50 ml each) of the following solutions: (1) ~ 80% of phenol in ethanol; (2) a mixture of water and pyridine (1:4), containing 1 ml of 002M NaCN; and (3) 6% of ninhydrin in ethanol. A little resin from the reaction vessel was transferred using a Schott microfilter (0.5 ml), the resin was washed with three portions of MeOH, transferred to a 3-ml tube, each of the above solution (by 3 drops) was added, and the tube was heated at 120 °C with shaking for 3–5 min.

The Boc protection was removed by treatment with 50% TFA for 0.5 h. The resin suspension foamed during reaction due to a vigorous gas evolution, Scheme 1.21:

Removal of the Boc Protection from Amino acids on the Merrifield Resin. A 55% solution of TFA in DCM (20 ml) was poured onto a 1 g sample of the resin with an immobilized amino acid placed in a high 30-ml vial. After the gas evolution was observed, the resin acquired a characteristic pinkish color. The vial was then shaken

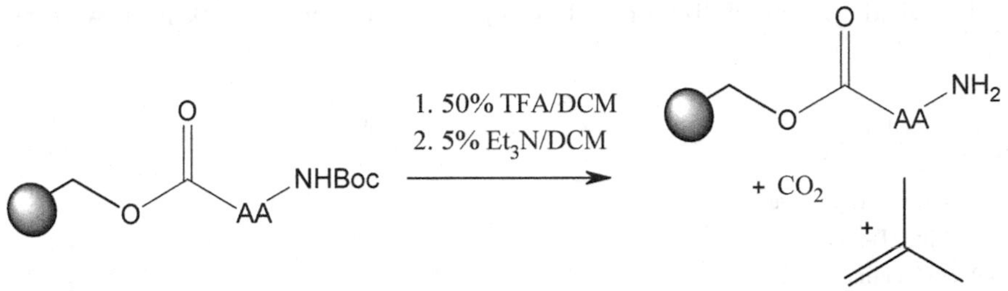

Scheme 1.21 Removal of Boc protection

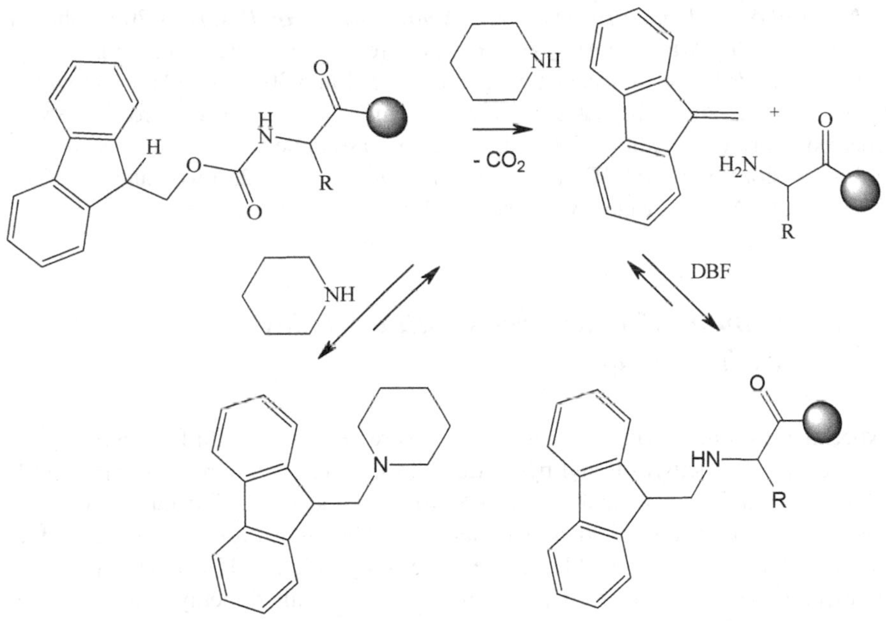

Scheme 1.22 Removal of Fmoc protection

for 0.5 h, and the resin was separated using a glass filter and washed with DCM (2 × 20 ml), 10% DIEA in DCM (2 × 30 ml), DMF (2 × 10 ml), DCM (15 ml), and MeOH (15 ml). The resin was dried in a vacuum oven and weighed. Amino groups were then detected by the Kaiser test.

The Fmoc protection from amino acids immobilized on the Rink support was removed selectively with a 20% solution of piperidine in DMF. The intermediate dibenzofulvene (DBF) rapidly reacts with excess piperidine to form 1-(9H-fluoren-9-ylmethyl)-piperidine which is water insoluble and precipitates when the reaction mixture is diluted with water (Scheme 1.22). The yield of immobilization of Fmoc-amino acids on the Rink resin is calculated from the weight of a dried 1-(9H-fluoren-9-ylmethyl) piperidine precipitate.

The yields of immobilization of Fmoc-amino acids on the Rink resin were as follows, %:

Glycine (Gly)	89
β-Alanine (β-Ala)	90
5-Aminovaleric (5-Ava)	84
L-Alanine (L-Ala)	94
L-Phenylalanine (L-Phe)	82
L-Proline (L-Pro)	86

Removal of the Fmoc Protection from Amino Groups on Resins. A 20% solution of piperidine in DMF (20 ml) was poured on a resin containing an Fmoc-protected amino group or Fmoc-protected amino acid, placed in a 30-ml standard vial with a porous bottom. The resin was separated from the solution, and the procedure was repeated; therewith, additional 30-min shaking was applied. The resin was washed in succession with DMF (3 × 30 ml), DCM (2 × 25 ml), and MeOH. (2 × 15 ml). Amino groups were detected by the Kaiser test.

1.3.3   Synthesis of Heterocycles with a Readily Leaving Group

Experiments on hetarylation of amino acids were performed with four fairly active heterocycles: 2-(methylsulfonyl)pyrimidine (**IV**), 2-bromo-5-nitro-1,3-thiazole (**V**), 2-fluoropyrimidine (**VI**), and 2-chloro-5-nitropyridine (**VII**). Pyrimidine **IV** was prepared in a moderate yield by oxidation of 2-(methylsulfonyl)pyrimidine (**III**) with *m*-chloroperbenzoic acid [15]. The starting pyrimidine **III** is readily prepared by methylation of commercial pyrimidine-2-thione with dimethyl sulfate in the presence of NaOH [16], Scheme 1.23.

2-Bromo-5-nitro-1,3-thiazole (**V**) was prepared by nitration of 2-amino-1,3-thiazole followed by the diazotization and introduction of bromine in position 2 by the Sandmeyer reaction [17] in Scheme 1.24. It should be added that the ^{1}H NMR spectrum is poorly informative and contains only one signal at δ 8.3 ppm. The IR spectrum (KBr) was more informative, and it contained the following

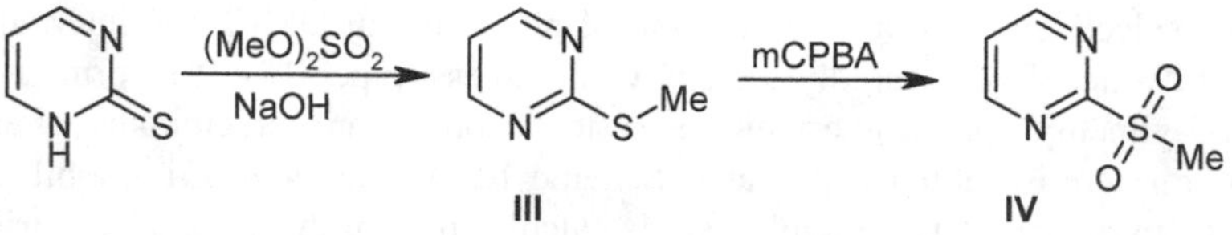

Scheme 1.23 Preparation of 2-(methylsulfonyl)pyrimidine (**IV**)

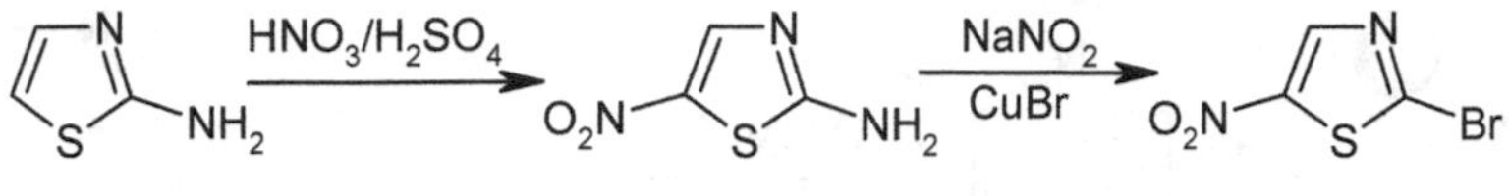

Scheme 1.24 Preparation of 2-bromo-5-nitrothiazole (**V**)

Scheme 1.25 Preparation of
2-fluoropyrimidine (**VI**)

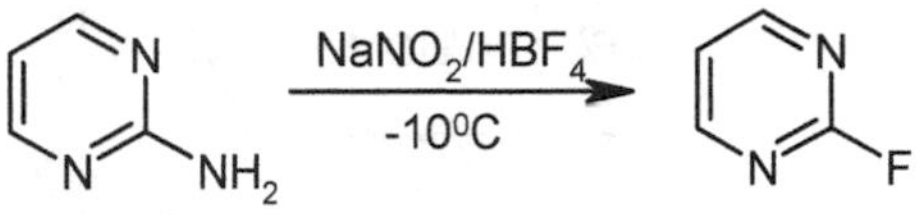

frequences: 3084 v (C–H), 1520 as, 1344 s (NO₂), 740 (C–Br). For comparison, the
IR spectrum of compound **V** from the site of Acros (www.acros.com) can be used.

2-Fluoropyrimidine (**VI**) was prepared by diazotization of 2-aminopyrimidine
under the action of sodium nitrite in 48% HBF₄ [10]. 2-Fluoropyrimidine is so
reactive that on attempted neutralization of the reaction mixture with 1 N NaOH by
the procedure [10], we always obtained a hydrolysis product, i.e., pyrimidin-2-one.
It was found that the yield of compound **VI** can be improved by using a suspension
of sodium hydrocarbonate instead of a solution of NaOH and cooling (to −10 °C)
during neutralization in Scheme 1.25.

2-Chloro-5-nitropyridine **VII**. A commercial reagent was used (synthesis of
compound **VII** is described in detail in [18]).

1.3.4 Realization of Stage 3 (Solid-Phase Nucleophilic Substitution) and Stage 4 (Product Removal from Polymer Support)

1.3.4.1 Synthesis of N-(Pyrimidin-2-yl) Amino Acid Derivatives

The nucleophilic substitution reactions with amino acids immobilized on the Rink
resin were performed using 2-fluoropyrimidine (**VI**) and DIEA as the base in
Scheme 1.26. During reaction, the resin volume decreased 1.5–2 times, which, too,
was evidence for reaction progress. We used the Kaiser test for amino group, taking
from the reaction mixture small resin portions as control of reaction progress. The
resulting primary amides of N-(pyrimidin-2-yl)-amino acids were removed from the
support by treatment with 50% TFA. Compounds **VIIIa–VIIIk** were purified by
column chromatography. The fundamental data analyses of compounds **VIII** were
satisfactory. Structural assessment was also performed by ¹H NMR spectroscopy
and mass spectrometry; the spectral data are listed in Table 1.3.

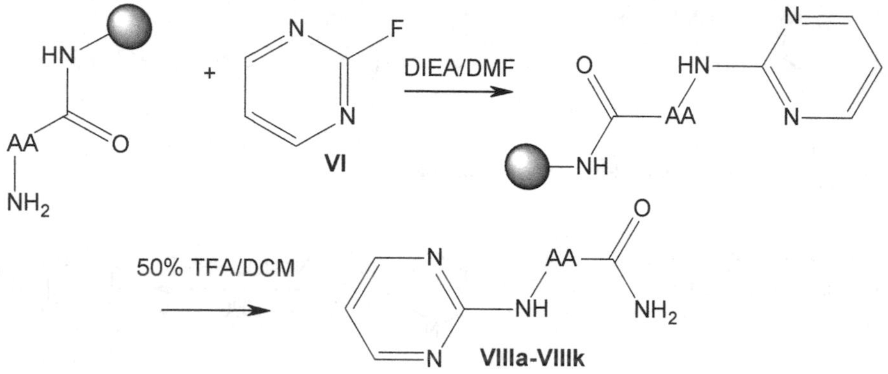

Scheme 1.26 Preparation of *N*-(pyrimidin-2-yl)amino acid amides

Synthesis of Amides of N-(pyrimidin-2-yl) Amino Acids. A solution of 0.52 (4 mmol) of DIEA in absolute DMF was poured onto 0.5 g of the Rink resin containing an unsubstituted amino acid. After 7–10 min, when the resin gained weight, 0.39 g (4 mmol) of 2-fluoropyrimidine was added. The vials were tightly closed and heated at 50–55 °C with intermittent shaking. The Kaiser test showed that the reaction was almost complete in 7–12 h. The reacted resin was successively washed using a filter with DMF (15 ml), DMF/DCM (15 ml 50%), DCM (2 × 25 ml), and MeOH (2 × 15 ml) and dried in a vacuum oven, after which it was transferred into a 30-ml flask, poured with 15 ml of 60% TFA/DCM, and agitated on an orbital shaker for 1 h. The resin was separated using a glass filter and washed with absolute DCM (15 ml) and MeOH (15 ml). The filtrate was evaporated on a rotary evaporator. The oily residue was poured with absolute MeOH (5 ml) and evaporated again. The procedure was repeated until TFA was removed completely. The product was purified, if necessary, by column chromatography (MeOH–CHCl₃, 4:1) on silica. TLC was performed in MeOH–CHCl₃, 4:1. The yield (Table 1.3) was calculated on the basis of the standard resin capacity.

We chose 2-(methylsulfonyl) pyrimidine (**IV**) (Scheme 1.27) for the activated pyrimidine for the reaction of activated pyrimidines with amino acids on the Merrifield resin. Reaction progress was followed by the Kaiser test for amino group. Removal of products from the resin was performed in the presence of sodium methylate. The resulting *N*-(pyrimidin-2-yl) amino acid esters **IXa–IXh** were characterized by the mass and ¹H NMR spectra (Table 1.3). Three examples were used to demonstrate the possibility to prepare free amino acids, viz. *N*-(pyrimidin-2-yl)amino acids **Xa–Xc,** by hydrolysis of the corresponding esters.

Synthesis of Methyl Esters of N-(pyrimidin-2-yl) Amino Acids (**X**): A solution (6 ml) of 0.65 (5 mmol) of DIEA in absolute DMF was poured into 1 g of the Merrifield resin containing an unprotected amino acid. After 5 min, 0.79 g (5 mmol) of pyrimidine **IV** was added to the resin. The vials were tightly closed and heated at 80–90 °C with intermittent shaking. Reaction progress was followed

Table 1.3 Yields and constants of N-(pyrimidin-2-yl)amino acid esters and their amides

Comp. no.	Compound	X	^{1}H NMR spectrum, δ, ppm 360 MHz, DMSO-d_6			Mass spectrum, M^+, m/z	Yield, %
			$H_{4,6}$	H_5	NH		
IXa		MeO	8.22	6.55	7.14	167	85
VIIIa	NHCH$_2$COX	NH$_2$	8.32	6.54	7.09	152	96
IXb		MeO	8.19	6.46	6.75	181	94
VIIIb	NH(CH$_2$)$_2$COX	NH$_2$	8.31	6.63	7.31	166	98
VIIIc	NH(CH$_2$)$_3$COX	NH$_2$	8.36	6.75	6.64	180	87
IXc		MeO	8.26	6.53	6.85	209	34
VIIId	NH(CH$_2$)$_4$COX	NH$_2$	8.28	6.56	6.47	194	65
IXd		MeO	8.22	6.54	7.14	182	79
VIIIe	NHCH(CH$_3$)COX	NH$_2$	8.33	6.54	7.25	166	88
IXe		MeO	8.25	6.53	6.66	217	32
VIIIf	NHCH(CH$_2$Ph)COX	NH$_2$	8.20	6.53	7.45	242	49
IXf		MeO	8.31	6.68	6.58	218	11
VIIIg	NHCH(Ph)CH$_2$COX	NH$_2$	8.19	6.48	6.58	242	23
VIIIh	NHCH(CH$_2$CONH$_2$)COX	NH$_2$	8.34	6.74	7.27	209	28
VIIIi	NHCH(i-Pr)COX	NH$_2$	8.25	6.57	6.81	195	35
IXg		MeO	8.25	6.71	–	208	52
VIIIj		NH$_2$	8.27	6.57	–	192	72
IXh		MeO	8.48	6.81	7.46	244	74
VIIIk	NHCH$_2$C$_6$H$_4$COX	NH$_2$	8.18	6.59	7.07	228	83

by the Kaiser test. The reacted resin was successively washed using a filter with DMF (25 ml), DCM (2 × 25 ml), and MeOH (2 × 15 ml), after which it was transferred into a standard 30-ml vial and poured with absolute THF (3 ml) and 2 M NaOMe in MeOH (1 ml). The vials were tightly closed and shaken at room temperature for 6 h. The resin was separated using a glass filter and washed with MeOH and THF. The filtrate was neutralized with 10% HCl and evaporated on a rotary evaporator. Methanol was used to treat the dry residue in order to extract the resulting ester. The extract was filtered off and evaporated on a rotary evaporator. When required, the product is purified, by column chromatography

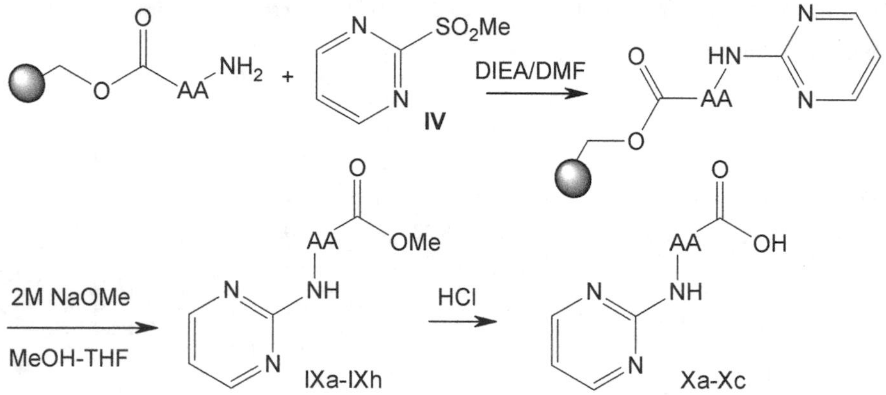

Scheme 1.27 Preparation of esters of *N*-(pyrimidin-2-yl)amino acids and their hydrolysis

(MeOH–CHCl$_3$, 2:9) on silica. The yield (Table 1.3) was calculated on the basis of the resin capacity. TLC was performed in MeOH–CHCl$_3$, 2:9.

Synthesis of Amino Acids **X** *from esters* **IX**. Ester **IX** (50–200 mg) was dissolved in 5 ml of 40% aqueous MeOH, after which 1 ml of conc. HCl was added. The solution was heated at 60 °C for 10–15 min and evaporated. The acid was recrystallized from aqueous methanol. TLC was performed in MeOH–CHCl$_3$– AcOH, 1:3:0.2. Acid **X** was prepared on the scale of 30–150 mg.

1.3.4.2 Synthesis of *N*-(5-Nitrothiazol-2-yl) Amino Acid Derivatives

Amino acids immobilized on the Rink and Merrifield resins were reacted with 2-bromo-5-nitrothiazole **V** in the presence of triethylamine in Scheme 1.28:

Synthesis of N-(5-nitrothiazol-2-yl) Amino Acid Amides. A solution (4 ml) of 0.2 (2 mmol) of triethylamine in absolute DMF was added to 0.5 g of the Rink resin containing an unprotected amino acid. After 10 min, 0.42 g (2 mmol) of 2-bromo-5-nitrothiazole was added. The vial was tightly closed and cooled to 5 °C

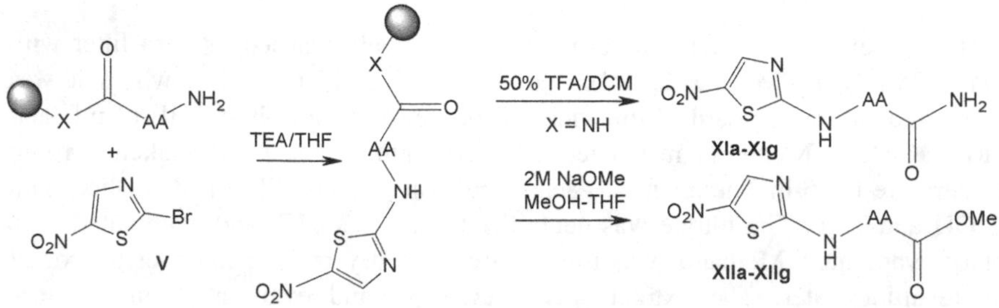

Scheme 1.28 Synthesis of esters and amides of *N*-(5-nitrothiazol-2-yl)amino acids

with constant shaking. The Kaiser test showed that the reaction is almost complete in 1–2 h. The reacted resin was washed continually using the filter with DMF (4 × 10 ml), DCM (2 × 15 ml), and MeOH (2 × 15 ml). After drying in a vacuum oven the resins were placed in a 30-ml vial, poured with 55% TFA/DCM (10 ml), and shaked on an orbital shaker for 1 h. The resin was separated using a glass filter and washed with absolute DCM (5 ml) and absolute MeOH (5 ml). The filtrate was evaporated on a rotary evaporator. The oily residue was poured with absolute MeOH (5 ml) and evaporated again. The procedure was repeated until TFA was removed completely. When necessary, the product was purified, by column chromatography (MeOH–CHCl₃, 1:2) on silica. The yield was calculated on the basis of the average resin capacity, in Table 1.4. Thus, we obtained amides **XIa–XIe, XIf,** and **XIg** (50–150 mg).

Synthesis of Methyl Esters of N-(5-nitrothiazol-2-yl)amino acids. A solution (3 ml) of 0.3 g (3 mmol) of triethylamine in absolute THF was added to 0.5 g of the Merrifield resin containing immobilized unprotected amino acids. After 10 min, a solution (3 ml) of 0.65 g (3 mmol) of 2-bromo-5-nitrothiazole in THF was added to the suspended resin. The reaction was performed in cold condition with intermittent shaking. The resin got intense red. After 10 h, the resin was separated using a glass filter, washed with DMF (2 × 20 ml), DCM (2 × 30 ml), and MeOH

Table 1.4 Yields and constants of *N*-(5-nitrothiazol-2-yl)amino acid amides **XI** and esters **XII**

Comp. no.	Compound	X	¹H NMR spectrum, δ, ppm		Mass spectrum [M⁺−NO₂], m/z	Yield, %
			H₄	NH		
XIIa	O₂N–[thiazole]–NHCH₂COX	MeO	8.07	7.25	172	81
XIa		NH₂	8.09	7.53	157	85
XIIb	O₂N–[thiazole]–NH(CH₂)₂COX	MeO	7.95	7.32	186	89
XIb		NH₂	7.94	7.56	171	76
XIc	O₂N–[thiazole]–NH(CH₂)₃COX	NH₂	8.12	7.37	184	79
XIIc	O₂N–[thiazole]–NH(CH₂)₄COX	MeO	7.91	7.34	214	92
		NH₂	–	–	198	–
XIId	O₂N–[thiazole]–NHCH(CH₃)COX	MeO	7.99	7.41	186	84
XId		NH₂	8.07	7.45	171	88
XIIe	O₂N–[thiazole]–NHCH(CH₂Ph)COX	MeO	7.90	–	262	38
XIe		NH₂	8.08	7.82	247	27
XIIf	O₂N–[thiazole]–NHCH(Ph)CH₂COX	MeO	7.92	–	262	52
–		NH₂	–	–	248	–
XIf	O₂N–[thiazole]–NHCH(i-Pr)COX	NH₂	8.01	7.89	185	22
XIIg	O₂N–[thiazole]–NHCH₂C₆H₄COX	MeO	7.95	7.37	248	64
XIg		NH₂	8.11	7.65	233	55

(2 × 15 ml), dried in a vacuum oven, and weighed. The dry resin was placed in a standard 30-ml vial and poured with absolute THF (3 ml) and 2 M NaOMe in MeOH (1 ml). The vial was tightly closed and shaken at a room temperature for 6–12 h. The resin was separated using a glass filter and washed with 30% MeOH in THF. The filtrate was neutralized with 10% HCl and evaporated on a rotary evaporator. The dry residue was treated with methanol to extract the resulting ester, and the extract was evaporated on a rotary evaporator. The product was recrystallized from hexane–ethyl acetate, 3:1, and dried in a vacuum oven. Thus, we obtained esters **XIIa–XIIg** (20–170 mg). An intense yellowish red coloration of the resin was observed with the resins. The Kaiser test showed that the reaction with 2-bromo-5-nitrothiazole occurs at a room temperature for 3–6 h. The resulting primary amides and methyl esters of N-(thiazol-2-yl) amino acids were characterized as the mass and ^{1}H NMR spectra. The characteristics of compounds **XIa–XIg** and **XIIa–XIIg** are listed in Table 1.4. The yields of chromatographic purity of the obtained N-(thiazol-2-yl) amino acid derivatives proved to be lower than those of N-(pyrimidin-2-yl) amino acid derivatives. We can therefore suggest that, along with the main nucleophilic substitution reaction, 2-bromo-5-nitrothiazole undergoes the side reaction involving thiazole ring opening (Scheme 1.29), as was mentioned in [19]:

1.3.4.3 Synthesis of N-(5-Nitropyridin-2-Yl) Amino Acid Derivatives

Commercial 2-chloro-5-nitropyridine (**VII**) and DIEA as the base were used to prepare N-(pyridin-2-yl) amino acid derivatives on the Rink resin in Scheme 1.30. The darkening of the support was observed during reaction. Reaction progress was followed by the Kaiser test for amino group. It then showed that the reaction occurs at 90–95 °C for 5–6 h, on average.

Synthesis of N-(5-nitropyridin-2-yl) Amino Acid Amides. A solution (4 ml) of 0.39 (3 mmol) of DIEA in absolute DMF was added to 0.5 g of the Rink resin containing an unprotected amino acid. After 10 min, 0.73 g (3 mmol) of 2-chloro-5-nitropyridine was added to the resin. The vial was tightly closed and heated at 95 °C for 12 h with intermittent shaking. The reacted resin was successively washed using a filter with DMF (4 × 10 ml), DCM (2 × 15 ml), and MeOH

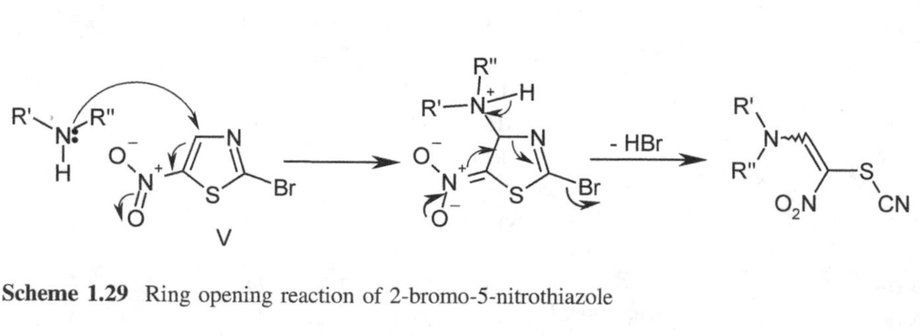

Scheme 1.29 Ring opening reaction of 2-bromo-5-nitrothiazole

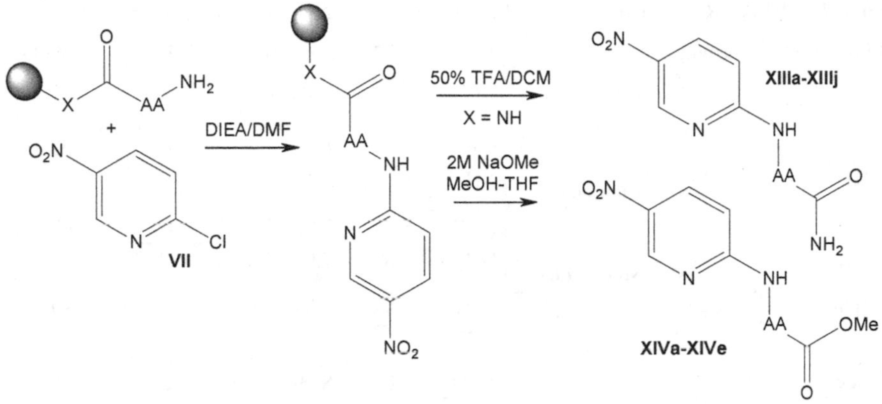

Scheme 1.30 Synthesis of esters and amides of N-(5-nitropyridin-2-yl)amino acids

(2 × 15 ml) and dried in a vacuum oven. The dry resin was placed in a 30-ml vial, poured with 10 ml of 50% TFA/DCM, and agitated on an orbital shaker for 1 h. The resin was separated using a glass filter and washed with absolute DCM (10 ml) and absolute MeOH (5 ml). The filtrate was evaporated on a rotary evaporator. The oily residue was poured with 5 ml of absolute MeOH and evaporated again. The procedure was repeated until TFA was removed completely. If necessary, the product was purified, by recrystallization from hexane–ethyl acetate. The yield was calculated on the basis of the average resin capacity (see Table 1.5). Thus, we prepared amides **XIIIa–XIIIg** (30–160 mg). Esters **XIV** were prepared in the same way on the Merrifield resin.

Synthesis of Methyl Esters of N-(5-nitropyridin-2-yl) amino acids: A solution of (4 ml) of 0.39 (3 mmol) of DIEA in absolute DMF was added to 0.5 g of the Merrifield resin containing an unprotected amino acid: 0.73 g (3 mmol) of 2-chloro-5-nitropyridine was added to the resin after 10 min. The vial was tightly closed and heated at 95 °C for 10–12 h with intermittent shaking. The reacted resin was successively washed using a filter with DMF (4 × 10 ml), DCM (2 × 15 ml), and MeOH (2 × 15 ml), dried in a vacuum oven, and weighed. The dry resin was placed in a standard 30-ml vial and poured with absolute THF (3 ml) and 2M NaOMe in MeOH (1 ml). The vial was tightly closed and shaken at room temperature for 5–7 h. The resin was separated using a glass filter and washed with 30% MeOH in THF, and the filtrate was neutralized with 10% HCl and evaporated on a rotary evaporator. The residue was treated with methanol to extract the product, and the extract was evaporated on a rotary evaporator. Thus, we obtained esters **XIVa–XIVe** (65–100 mg). Compounds **XIIIa–XIIIg** and **XIVa–XIVe** were characterized by the ^{1}H NMR spectra (Table 1.5).

Table 1.5 ^{1}H NMR data and constants of N-(5-nitropyridin-2-yl)amino acid derivatives

Comp. no.	Formula	X	^{1}H NMR spectrum, δ, ppm			Purity[a]	Yield, %
			H_6	H_4	H_3		
XIVa	O_2N ... N NHCH$_2$COX	MeO	8.87	8.15	6.53	B	57
XIIIa		NH_2	8.87	8.16	6.77	B	72
XIVb	O_2N ... N NH(CH$_2$)$_2$COX	MeO	8.84	7.96	6.57	C	68
XIIIb		NH_2	8.82	7.98	6.55	C	60
XIIIc	O_2N ... N NH(CH$_2$)$_3$COX	NH_2	8.83	7.98	6.51	C	57
XIVc	O_2N ... N NH(CH$_2$)$_4$COX	MeO	8.82	8.09	6.61	C	70
XIIId		NH_2	8.83	7.97	6.54	C	47
XIVd	O_2N ... N NHCH(CH$_3$)COX	MeO	8.80	8.14	6.65	B	52
XIIIe		NH_2	8.83	7.97	6.62	C	84
XIIIf	O_2N ... N NHCH(CH$_2$Ph)COX	NH_2	8.80	8.16	6.67	B	28
XIIIg	O_2N ... N NHCH(Ph)CH$_2$COX	NH_2	8.81	7.97	6.58	C	18
XIIIh	O_2N ... N NHCH(i-Pr)COX	NH_2	8.83	8.03	6.81	B	12
XIIIi	O_2N ... N (pyrrolidine) O-X	NH_2	8.88	8.12	6.78	B	32
XIVe	O_2N ... N NHCH$_2$C$_6$H$_4$COX	MeO	8.85	8.14	6.55	A	69
XIIIj		NH_2	8.84	8.02	6.59	A	39

[a]The purity of products **XIIIa–XIIIg** and **XIVa–XIVe** was estimated by the ^{1}H NMR spectra as follows: A (<5% of admixtures), B (<10% of admixtures), and C (>20% of admixtures)

1.4   Conclusions

The 4-stage solid-phase reaction procedure for the preparation of non-natural amino acids, optimized by us in a student's laboratory course, is quite convenient from the methodical viewpoint. No special equipment is necessary, the yields of the target products are high, and the isolated quantities are sufficient for full-scale characterization.

The reactions can be accomplished concurrently with variable number of amino acids and hetarylating agents (e.g., in the 2 × 2, 2 × 3, or 3 × 3 formats). Amino acids themselves (unlike amino acids immobilized on resins) are commercially available and inexpensive. As a result of this, the total number of stages in a training task should be determined by the possibilities of the training laboratory. We set out for ourselves the task to work out the *total cycle* (including protection and immobilization stages).

References

1. Ermolat'ev DS, Babaev EV (2003) Solid-phase synthesis of methyl N-(pyrimidin-2-yl) glycinate. Molecules 8:467–471. doi:10.3390/80600467
2. Ermolat'ev DS, Babaev EV (2005) Solid-phase synthesis of N-(pyrimidin-2-yl)amino acids amides. ARKIVOC 172–178. http://www.arkat-usa.org/get-file/20198/
3. Larsen SD, Connell MA, Cudahy MM, Evans BR, May PD, Meglasson MD, O'Sullivan TJ, Schostarez HJ, Sih JC, Stevens FC, Tanis SP, Tegley CM, Tucker JA, Vaillancourt VA, Vidmar TJ, Watt W, Yu JH (2001) Synthesis and biological activity of analogues of the antidiabetic/antiobesity agent 3-guanidinopropionic acid: discovery of a novel aminoguanidinoacetic acid antidiabetic agent. J Med Chem 44:1217–1230. doi:10.1021/jm000095f
4. Vailancourt VA, Larsen SD, Tanis SP, Burr JE, Connell MA, Cudahy MM, Evans BR, Fisher PV, May PD, Meglasson MD, Robinson DD, Stevens FC, Tucker JA, Vidmar TJ, Yu JH (2001) Synthesis and biological activity of aminoguanidine and diaminoguanidine analogues of the antidiabetic/antiobesity agent 3-guanidinopropionic acid. J Med Chem 44:1231–1248. doi:10.1021/jm000094n
5. Nicolau KC, Hanko R, Hartwig W (2002) Handbook of combinatorial chemistry. Wiley–VCH, Weinheim
6. Iwanowicz EJ, Poss MA, Lin J (1993) Preparation of N, N'-bis-tert-Butoxycarbonylthiourea. Synth Commun 23:1443–1445. doi:10.1080/00397919308011234
7. Carpino LA, Grace YH (1972) 9-Fluorenylmethoxycarbonyl amino-protecting group. J Org Chem 37:3404–3409. doi:10.1021/jo00795a005
8. Kaiser E, Colescott RL, Bossinger CD, Cook PI (1970) Color test for detection of free terminal amino groups in the solid-phase synthesis of peptides. Anal Biochem 34:595–611. doi:10.1016/0003-2697(70)90146-6
9. Brown DJ, Waring PJ (1974) Pyrimidine reactions. Part XXV. Synthesis and piperidinolysis of some simple fluoropyrimidines. J Chem Soc Perkin Trans 2:204–208. doi:10.1039/P29740000204
10. Brown DJ, Ford PW (1967) Simple pyrimidines. Part X. The formation and reactivity of 2-, 4-, and 5-pyrimidinyl sulphones and sulphoxides. J Chem Soc C 568–572. doi:10.1039/J39670000568
11. James IW (1999) Linkers for solid phase organic synthesis. Tetrahedron 55:4855–4946. doi:10.1016/S0040-4020(99)00125-8
12. Kobayashi S, Akiyama R, Kitagawa H (2001) Polymer-supported glyoxylate and alpha-imino acetates. Versatile reagents for the synthesis of alpha-hydroxycarboxylic acid and alpha-amino acid libraries. J Comb Chem 3(2):196–204. doi:10.1021/cc0000850
13. Tarbell DS, Yamomoto Y, Pope BM (1972) New method to prepare N-t-butoxycarbonyl derivatives and the corresponding sulfur analogs from di-t-Butyl dicarbonate or di-t-Butyl dithiol dicarbonates and amino acids. Proc Natl Acad Sci USA 69:730–732
14. Greene TW, Peter GM (1999) Protective groups in organic synthesis. Wiley–VCH, Weinheim
15. Adlington RM, Baldwin JE, Catterick D, Pritchard GJ (1999) The synthesis of pyrimidin-4-yl substituted α-amino acids. A versatile approach from alkynyl ketones. J Chem Soc Perkin Trans 1:855–866. doi:10.1039/A806741D
16. Boarland MPV, McOmie JFW (1952) Pyrimidines. Part II. The ultraviolet absorption spectra of some monosubstituted pyrimidines. J Chem Soc 3716–3722. doi:10.1039/JR9520003716
17. Deshpande RN, Nargund KS (1974) Aryl 5-Nitro-2-thiazolyl sulphides, sulphones and ethers as potential antibacterials. J Prakt Chem 316(2):349–352. doi:10.1002/prac.19743160223
18. Chichibabin AE (1914) Zh Ross Fiz-Khim O-va 46(2):1236–1296
19. Ilvespaa AO (1968) Ringöffnung von nitrothiazolen. Helv Chim Acta 51(7):1723–1733. doi:10.1002/hlca.19680510729

Chapter 2
The Choice of Tools for Implementing Multistage Transformations. SPOS for Beginners

In Chap. 1, we remembered the ABC of solid-phase synthesis and dwelt in detail on the experimental problems associated with this technique. The principal inconvenience in the practical work with polystyrene resins is their tendency to swell in solvents. Solvents (and reagents) penetrate into the polymer globules, resulting in sharply increasing their volume. The gelatin-like nature of the swollen polymer makes it difficult to transfer from one vial to another or to a filter. In the present paper, we analyze the approaches to solving this problem. Three training tasks tested at the special student laboratory of combinatorial chemistry at the Moscow State University (MSU) were used as examples. In this work, we acquired a certain experience which can be useful for teachers and synthetic organic chemists.

Below are the three examples of tools facilitating manipulations with polymer supports we considered. In the first example, some problems are approached by the use of a container for resins and the Bill-Board kit in which reaction vials are combined with filters. The chemical transformation is represented by double modification of a resin-immobilized protected glycine (consecutive C-alkylation and N-acylation). In the second example (the tea-bag technique), a resin is contained in a porous plastic bag, and the chemistry is presented by a multistage sequence of transformations. In the third example, we consider the methodology of operation with specially modified polypropylene pins (lanterns), a new "synthesis philosophy" which radically changes all traditional views on modern organic synthesis. This technique was employed in a multistage (5–6 stages) synthesis of thioureas and guanidines. All the three examples represent versions of training tasks.

© The Author(s) 2017
E.V. Babaev, *Incorporation of Heterocycles into Combinatorial Chemistry*,
SpringerBriefs in Molecular Science, DOI 10.1007/978-3-319-50015-7_2

2.1   Bill-Board Technique

Let us consider a four-stage modification of glycine (containing benzophenone protection) immobilized on the Wang resin. The main feature of the task is the use of a simple technical tool, the Bill-Board kit, which is a specially designed set of filter tubes and convenient plastic facilities for stirring and filtering in several reaction vials. This is an inexpensive kit (if desired, its analog can be readily made), which allows one to facilitate and speed-up manipulations when performing con-current solid-phase reactions. The suggested task had been tested by a group of MSU students in the special laboratory course on combinatorial chemistry few years ago. To fulfill the task, three laboratory lessons and a final seminar for purity analysis of the synthesized products (lesson 4) are required.

2.1.1   Chemical Aspect of the Task

The possibility of C-alkylation of N-substituted amino acids immobilized on a support had first been demonstrated almost 20 years ago [1] (Scheme 2.1). The amino group in the resin-immobilized glycine is protected with the benzophenone fragment. The CH_2 group of the resulting imine acquires sufficient acidity char-acteristic of aza-allyl systems. Therefore, under the action of a strong base, for example, a substituted phosphazene (BEMP), it can be deprotonated and then C-alkylated.

Ever since the first publications, this line of research has substantially extended: Scott et al. [2] have published a big series of papers on the methodology of solid-phase synthesis by this reaction. Analogous reaction could be performed with low-active alkyl halides, as well as with alkenes (Michael reaction). So, different

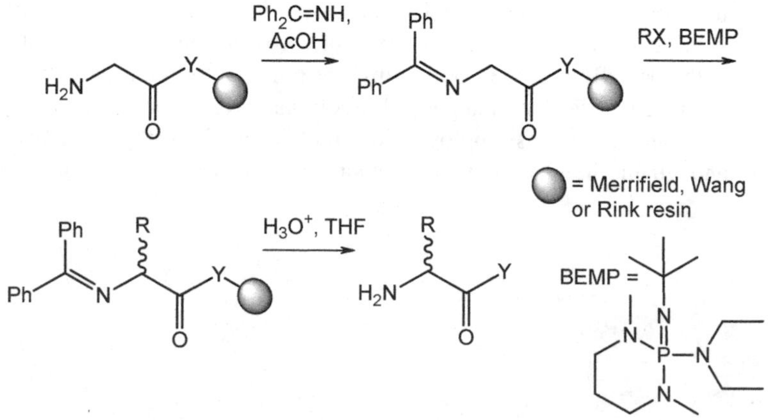

Scheme 2.1 Step-wise protocol of C-alkylation of protected amino acids on solid support

classes of compounds can be prepared: amides with the Rink resin and aminoaldehydes and aminoketones with the Weinreb resin; it all depends on the nature of the resin. It was found that using strong bases allows preparation of C-dialkylated amino acids. The resulting immobilized intermediates can be reacted with nucleophiles or converted into more complex acyclic and cyclic systems, for example, proline derivatives, lactams, and hydantoins.

2.1.2 Specific Features of the Bill-Board Kit

The facility and reproducibility of chemical transformations involving immobilized benzophenone imine derivatives of amino acids prompted Prof. Scott to adapt these sequences for the students' laboratory course in solid-phase synthesis at the Indiana University (USA). The Bill-Board training kit (Fig. 2.1) was constructed to facilitate concurrent manipulations with resins.

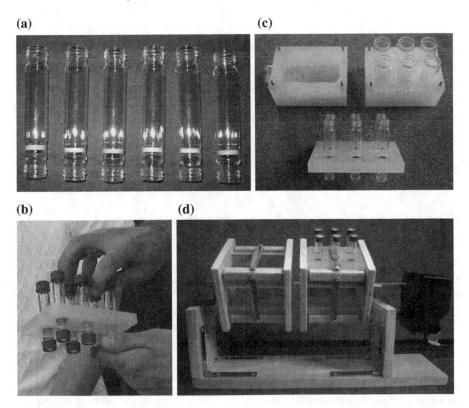

Fig. 2.1 Bill-Board kit: **a** double-ended filter vials; **b** fixing of reaction vials; and **c** additional plastic collectors: (*left*) for washings, (*right*) for final products, and (*bottom*) Bill-Board ready for fixing; and **d** option for simultaneous agitation of six reaction vials. A simple facility allows to fix eight Bill-Boards and simultaneously agitate 48 vials

Fig. 2.2 Features of practical implementation of the task: **a** addition of reagents at the first stage ▶ (lower caps are closed); **b** pump purging of resins on filters; **c** view of a closed Bill-Board during reaction; **d** removal of products from resins (under hood!); **e, f** different variants of mixing in 6 reaction vials

The Bill-Board is a specially designed set of filter tubes (Fig. 2.1a) with double-sided plastic screw caps with Teflon septa (Fig. 2.1b). As a result, each vial plays the role of either a reaction vessel (when the caps are screwed), or a filter (when the caps are removed). The filter is soldered unsymmetrically, and the solid support is placed in a larger vial compartment. The tubes are fixed in a special plastic holder (Bill-Board as such, Fig. 2.1b and c) which can be used both as a stand (on adding reagents, washing tube contents, and filtering) and as a block for simultaneously rotating and agitating six vials. Two convenient plastic accessories (Fig. 2.1c) serve as receivers for washings or stands for collection of final cleavage products, respectively. However, one can use simple accessories for agitating several Bill-Boards (Fig. 2.1d, see also Fig. 2.2e and f).[1]

2.1.3　A Brief Description of the Task

The task involves four stages: (1) alkylation of the CH_2 group of an amino acid immobilized on the Wang resin (W); (2) removal of the protective group; (3) acylation of the NH_2 group of the amino acid; and (4) removal of the doubly modified amino acid from the support (Scheme 2.2).

Each student performed six syntheses simultaneously, using three benzylating and two acylating agents, in a single Bill-Board labeled according to Scheme 2.3 (A, B and 1, 2, 3).

2.1.4　Reagents Used

The reagents used are as follows: resin (protected glycine on the Wang resin) 2 g, phosphazene base BEMP (Aldrich-79,432, see the formula in Scheme 2.1) 5 ml, diisopropylethylamine (DIEA) 100 ml, TFA (99%) 100 g, and aqueous HCl (1 N) 50 ml. Solvents are as follows: 1 l of N-methylpyrrolidin-2-one (NMP) and by

---

[1]A complete synthetic kit costs 175 USD, and it is designed for multiple uses and allows 6 concurrent combinatorial solid-phase reactions. The simplicity and low cost of the kit also allow successful solid-phase combinatorial syntheses to be performed by degree and PhD students not only at the Indiana University, but also at the Universities of Barcelona (Spain) and Lublin (Poland). In 2005, the Bill-Board was acquired by the special students' laboratory of combinatorial chemistry at the MSU, and in the framework of Prof. Scott's visit to Moscow, we practically implemented the task with a group of degree and PhD students of the Chemical Department.

(a)   **(b)**

(c)   **(d)**

(e)   **(f)**

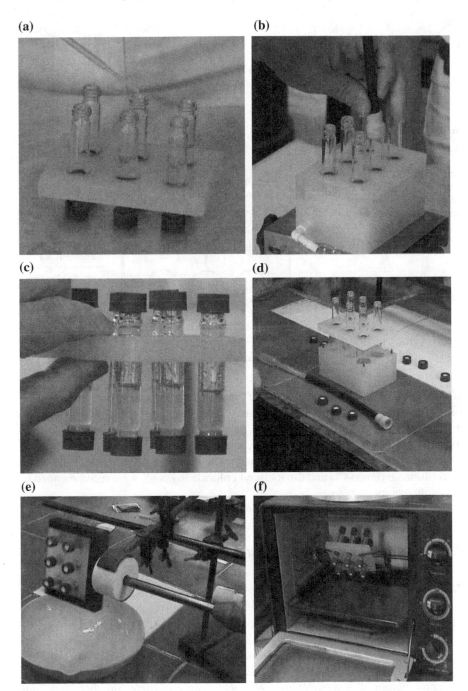

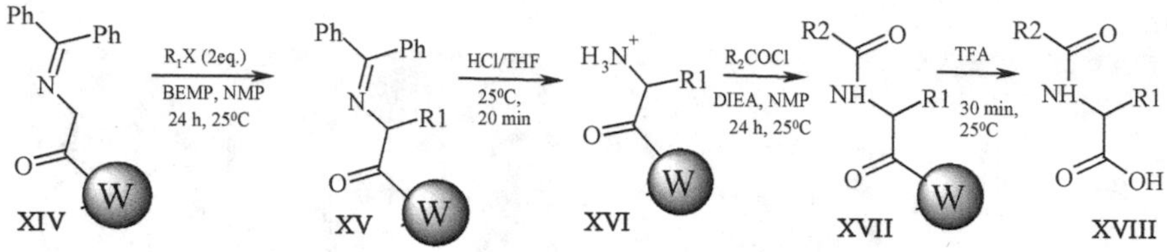

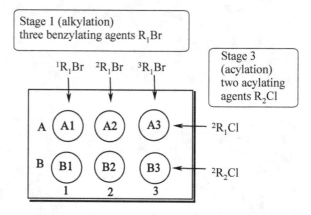

Scheme 2.2 Overview of the task (see text)

Scheme 2.3 6 samples of alkylated and acylated amino acid to be synthesized by every student

Stage 1 (alkylation)
three benzylating agents R_1Br

Stage 3 (acylation)
two acylating agents R_2Cl

^1R_1Br ^2R_1Br ^3R_1Br

A   (A1)   (A2)   (A3)   ←── ^2R_1Cl

B   (B1)   (B2)   (B3)   ←── ^2R_2Cl

   1   2   3

100 ml of dichloromethane (DCM), THF, and DMF (the latter is needed for rinsing caps of reaction vials). The alkylating agents were three benzyl bromides (from 1 to 5 g): unsubstituted benzyl bromide and its *m*-bromo and *p*-trifluoromethyl derivatives. (At the MSU, we used additionally a pyridine analog, specifically 2-chloro-5-(chloromethyl) pyridine.) Compounds with a high molecular weights, 9-fluorenyl chloroformate (Fmoc-Cl) and 2-naphthoyl chloride (by 5 g), were recommended as acylating agents.

2.1.5   Implementation of the Task

2.1.5.1   Lesson 1. *Alkylation* (Scheme 2.4)

(1) Preparation of an isopycnic solution of glycine imine **XIV** immobilized on the Wang resin, based on 65 mg (50 mmol) per one experiment (resin capacity 0.77 mmol g^{-1}). (*Isopycnic solution*: *two solvents with different densities, for example, THF and DCM, are added to a dry resin. By varying their ratio, one can obtain a homogeneous gel, when resin particles neither float nor precipitate*).

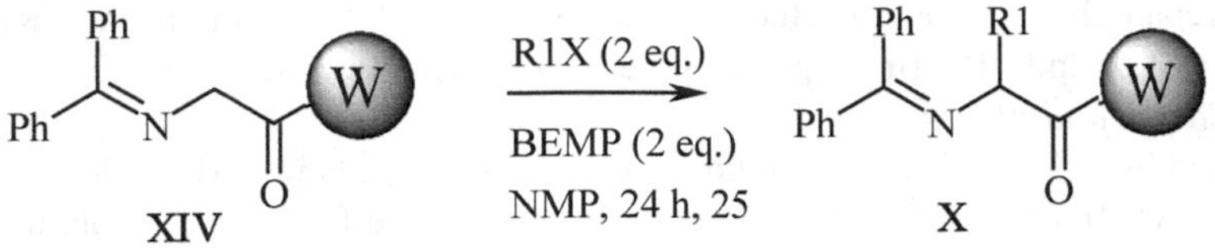

Scheme 2.4 C-alkylation of protected glycine immobilized on Wang resin

At the beginning of the first laboratory lesson, students get 6 reaction vials placed in a Bill-Board and a solution of resin **XIV** (Fig. 2.2a). However, the effectiveness of the method of equal distribution of the resin—in the form of suspension—among several tens of reaction vials should be mentioned (imagine a time required for weighting into each vial 65 mg of a dry resin!).

(2) The required volume of the isopycnic solution is distributed between vials, after which the solvent is allowed to drain. To expel residual solvent, a simple pump can be used. The resin in each reaction vessel is washed with NMP (3×3 ml) by means of a 3.5-ml plastic Beral pipette. At all stages of the work, one should meet the following instructions:

I. *General Washing Procedure*: *Washings should always be performed in a strictly specified order. The washing liquid, ~3 ml (~80% the total volume of the reaction vial), is added with a 3.5-ml pipette. Wait for 30 s to let the solvent to drain through the filter under the gravitation force. Then, the resin should be dried completely (pump purging) (Fig. 2.2b). A glass should be put beneath the drain pan plug. Finally, the liquid in the glass should be poured into a waste container.*

II. *The Use of Air Pump*: *The "pump" is a piece of a rubber tube with one end of each plugged with a plastic Beral pipette and the second end attached to septa through a slot. When the solvent has drained through the filter (within 30 s), attach the pump (from the septa side) to the vial neck and press the extended part of the pipette. To avoid back soaking of the solvent from the vial to tube, raise slightly the septa over the neck and only then unfold fingers. Repeat the procedure until all solvent drains through the filter.*

III. *Manipulations with Caps of Reaction Vials*: *Before screwing the cap, remove solvent residues from the space between the cap and vial by means of a soft cloth. Before unscrewing, turn the side of the Bill-Board to be opened up. Unscrew the cap, turn the Bill-Board around, and put it into the pan. Shake Bill-Board several times to make sure that all gel residues are removed from inner side of the cap. Open the top caps of all reaction vials, and put them into a glass for further use.*

(3) Mount the Bill-Board to a special holder. Screw the bottom caps (closest to soldered-in filters) of each reaction vial (Fig. 2.2c).

(4) Prepare three calibration Beral pipettes for adding benzylating agents $^1R_1$–Br, $^2R_1$–Br, and $^3R_1$–Br (separate pipette for each reagent). *Safety measures: glasses and gloves!*

(5) Add by 0.5 ml of 0.2 M solutions of alkylating agents in NMP (100 mmol, 2 equiv). $^1R_1$–Br is added to the reaction vials in the first *vertical* column (A1 and B1); $^2R_1$–Br, to the vials in the second vertical column (A2 and B2); and $^3R_1$–Br, to A3 and B3, respectively (Scheme 2.3). Add by 0.5 ml of a 0.2 M solution of BEMP in NMP (100 mmol, 2 equiv) in each of the six reaction vials. Screw the top caps of the Bill-Board, and mount it to a rotating apparatus (Fig. 2.2e and f). Record reaction initiation time and the Bill-Board number. Reaction time is 24 h. These reactions (and all subsequent reactions) are performed at a room temperature. Before leaving the laboratory, wash the drain tray with acetone over the waste container.

2.1.5.2   Lesson 2. Protection Removal and N-Acylation

(6) Record reaction completion time and take the Bill-Board out of the rotating apparatus. Unscrew the caps as recommended in Instruction III.

(7) Wash out excess reagents from the resin. Wash the alkylated resin (product **XV**) one time with 3 ml of THF and pump-dry (Instruction II).

(8) Mount the Bill-Board on a holder. Take 12 clean caps, screw the bottom caps of each reaction vial, and add about 2.5 ml of a 1 N aqueous solution of a 1:2 HCl–THF mixture to each vial. Screw the top caps, and place in the rotating apparatus for 20 min (Scheme 2.5).

(9) Unscrew all caps.

(10) Filter off and wash product **XVI** in succession with 3 ml of THF and 3 ml of NMP (one time).

(11) Mount the Bill-Board on a holder. Take 12 clean caps. Screw the bottom caps of each reaction vial, and add 0.5 ml of a 0.2 M solution of the first acylating agent $^1R_2$–COCl in NMP (100 mmol, 2 equiv) to resin **XVI** in the three reaction vials in the first horizontal row (A1, A2, and A3). Then, add 0.5 ml of a 0.2 M solution of the second acylating agent $^2R_2$–COCl in NMP (100 mmol, 2 equiv) to resin **XVI** in the three reaction vials in the second

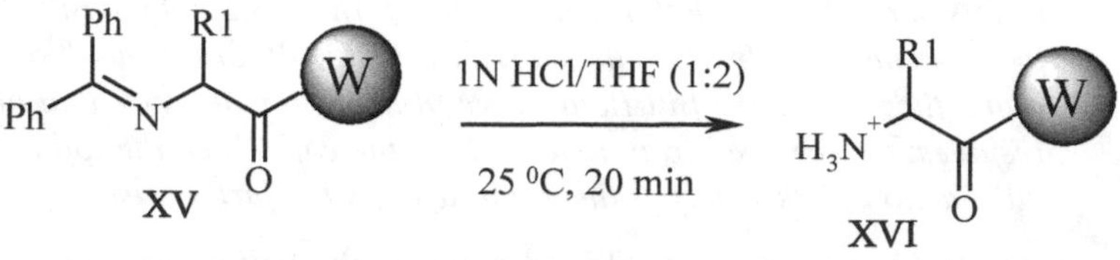

Scheme 2.5 Acidic deprotection of C-alkylated glycine immobilized on Wang resin

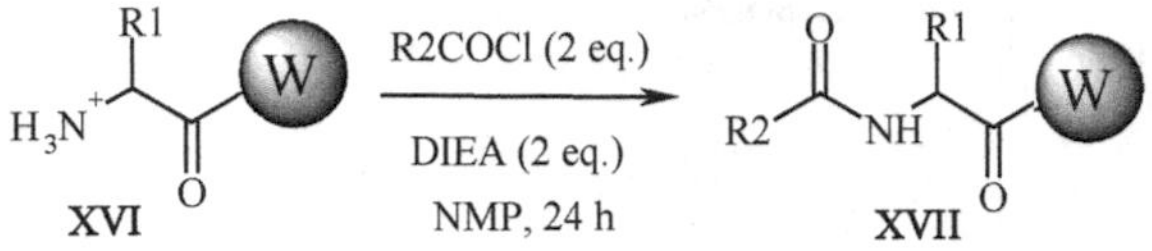

Scheme 2.6 N-acylation of C-alkylated glycine immobilized on Wang resin

horizontal row (B1, B2, and B3). Then, add 0.5 ml of a 0.3 M solution of DIEA in NMP (150 mmol, 3 equiv) to each of the six reaction vials. Screw the top caps. The reaction is performed for 24 h in the rotating apparatus (Scheme 2.6).

2.1.5.3 Lesson 3. Removal of N,C-Substituted Amino Acid from Support

(12) Take the Bill-Board out of the rotating apparatus and unscrew caps (Fig. 1.2d).

(13) Filter off and wash product **IV** with NMP (2 × 3 ml), THF (2 × 3 ml), and CH$_2$Cl$_2$ (3 × 3 ml).

(14) Mount the Bill-Board on a holder. Screw the bottom end with *clean* caps. Add 2 ml of CF$_3$COOH/H$_2$O (95:5) to each vial. *Take caution!* Screw the top caps and place the Bill-Board in a rotating apparatus for 30 min (Scheme 2.7).

(15) As the cleavage reaction is in progress, prepare six weighed receiver vials, label them (A1B1, A1B2, etc.), and place in a 6-hole rack.

(16) Products **XVIII** which are now in a *solution* (therefore, one should preserve the *filtrate*) transfer to the vials. To this end, turn the Bill-Board over (the bottom caps up), unscrew these caps, and mount receivers on the reaction vials. Make sure that the labels on the reaction and receiver vials are the same. Having accommodated all vials, mount the rack on them and *turn over the whole construction.* Unscrew the upper caps, and collect the filtrate into vials, using a pump.

(17) Wash the resins one time with 2 ml of CF$_3$COOH–H$_2$O (95:5) and one time with 2 ml of CH$_2$Cl$_2$, collecting washings into receiver vials. Every time, pump the solvent thoroughly down.

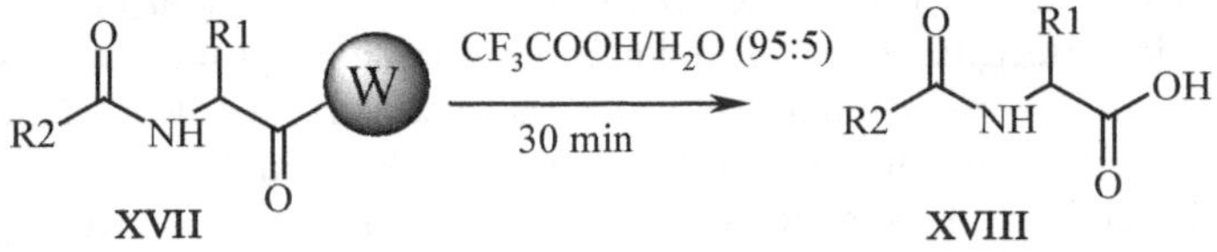

Scheme 2.7 Removal of the amino acid from Wang resin

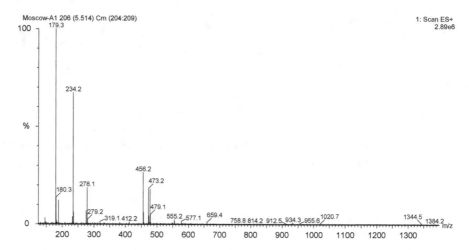

Fig. 2.3 Example LC-MS data for sample A1 (purity 95%, calculated for $C_{25}H_{20}F_3NO_4$: M 455)

(18)   Take the Bill-Board off from the vials. Transfer by 0.1 ml of each sample of product **V** into vials for subsequent LC/MS analysis (Fig. 2.3). The vials with the final products should be transferred to a vacuum drying oven for evaporation. (Evaporation can also be performed on a rotary evaporator in weighed flasks.)

(19)   Wash the drain pan and Bill-Board with acetone.

2.1.5.4   Lesson 4. Determination of Product Yields and TLC Analysis

(20)   Weight all final products **XVIII**. (Taking into account that the reactions are performed on a 50 μmol scale, and the molecular weights of products vary in the range 300–400 g mol^{-1}, the theoretical yield in each reaction should be 15–20 mg). The real yield is 10–15 mg.

(21)   For TLC, dissolve each product in THF (about 0.1 ml of THF per 1 mg of product) and use the system $CHCl_3$/THF/CH_3COOH (85/15/2). Develop the chromatograms first under UV light and then with iodine. Record R_f in both cases. As references, use synthesized samples mentioned earlier.

The results obtained in the same combinatorial task in several universities all over the world formed the basis for quite an unusual concept named "Distributed Drug Discovery." Its essence consists in the fact that students of different universities fulfilling the same training task obtain eventually a fairly representative combinatorial library of potential drugs. Synthesis and subsequent testing are planned from a united center [3]. The contribution of the Moscow's team to this project is described in a separate paper [4].

2.2 Tea-Bag Technique

This technique is a result of a long-term collaboration between the Chemical Diversity Research Institute (CDRI, Khimki) and Higher Educational Institutions of Moscow. The laboratory work was developed based on the practical experience in solid-phase synthesis, accumulated at the CDRI. Some students of the MSU involved in fulfilling the tasks of the special laboratory course in combinatorial chemistry could get familiar with the CDRI's procedures during excursions to this institute or Open Days (held since 2004). We have described the procedure of a student training task for such special laboratory courses below. The laboratory work is based on a technique widely used in solid-phase synthesis and called the tea-bag technique. This technique is known since 1985 due to the pioneering work of R. Houghten, a chemist from Scripps, La Jolla, who applied it to multiple peptide synthesis [5].

The container for resin here is a bag of a porous polypropylene film sealed over the entire perimeter (Fig. 2.4), for example, using a simple hot cutter (Fig. 2.4a). Such bags (Fig. 2.4b) can be easily made and put in a great number of vials to add similar reagents (Fig. 2.4c) or combined and put in one vial (Fig. 2.4d) to add the same reagent or at repeated resin washing stages. Naturally, the resin swelling effect is unavoidable, but here, one can fully avoid resin sticking to glass vial walls. Moreover, bags can function as a vessel and a filter simultaneously.

2.2.1 Reaction Sequence

As a model chemical reaction, we chose a sequence shown in Scheme 2.8. 2-Fluoro-5-nitrobenzoic acid, a compound containing a halogen activated for nucleophilic substitution and a nitro group which can be readily reduced to form an amino function and then acylated by the latter, is immobilized on the Wang resin. Theoretically, the starting acid can be a three-point template, since, for example, the third function (carboxyl), too, can be easily converted into an amide one. However, in the task under consideration, a shortened sequence is used, and the target products are substituted 2,5-diaminobenzoic acids **XXVII**.

Required Reagents and Equipment. The Wang resin, film for bags, and instrumentation for fabrication of the latter are required, as well as a large volume of different solvents, a series of secondary aliphatic amines, chlorides of aromatic (including heteroaromatic) carboxylic acids, $SnCl_2$, pyridine, acetic anhydride, and trifluoroacetic acid. The work is performed on a shaker (mostly at room temperature). Varied-size capped vials are used as reservoirs for bags with resins.

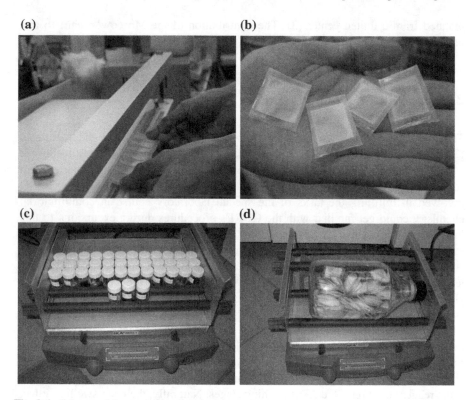

Fig. 2.4 Scheme of the tea-bag technique (see text)

2.2.2   *Implementation of the Work*

Stage 1. *Immobilization of 2-fluoro-5-nitrobenzoic acid on the Wang resin*: Wang resin **XX** (50 g, capacity 2.0 0.77 mmol g^{-1}) is suspended in a solution of 0.2 mol of 2-fluoro-5-nitrobenzoic acid **XIX** (37.41 g) and 25 mg of 4-(dimethylamino) pyridine in 400 ml of absolute dichloromethane (DCM), after which 31.5 ml (0.2 mol) of diisopropylcarbodiimide was added with caution (Scheme 2.9). The reaction mixture is shaken for 2 days at a room temperature. The resin is filtered off and washed with solvents in the following order: 2 × DCM, DMF, MeOH, 2 × DCM, 2 × MeOH, 2 × DCM, and 2 × hexane, and then vacuum-dried. Resin **XXI** is ready for further use.

    Stage 2. *Substitution of fluorine by dialkylamino group*: Resin **XXI** is distributed over plastic bags (Fig. 2.4a, b) whose number is equal to the number of final products. The resin weight is 250 mg (capacity 1.5 mmol g^{-1}). According to the statements of work (which specify what amine to take and how many experiments to perform), the bags are distributed over vials equal in number to the number of amines **XXII** (Fig. 2.4c and Scheme 2.10).

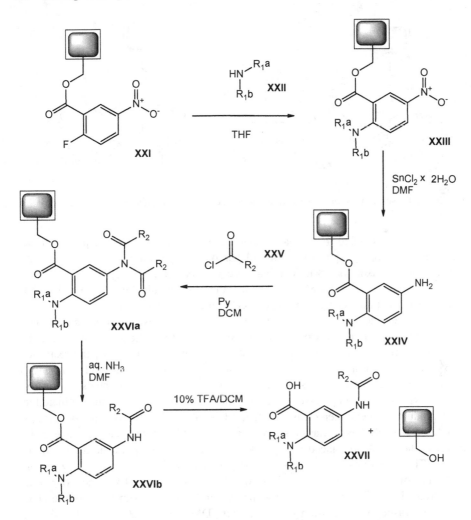

Scheme 2.8 Overview of the plan of using the tea-bag technique

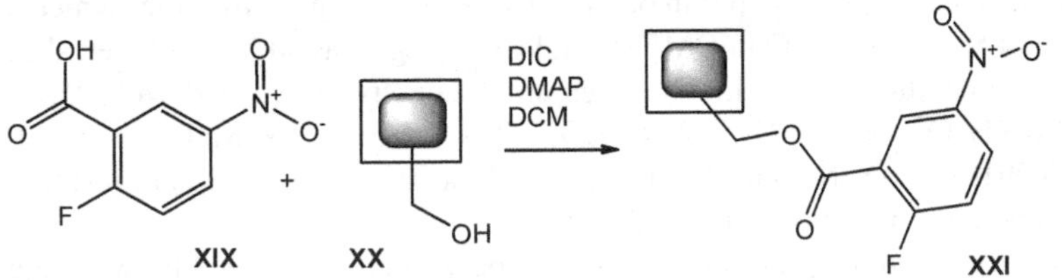

Scheme 2.9 Immobilization of aromatic acid on Wang resin

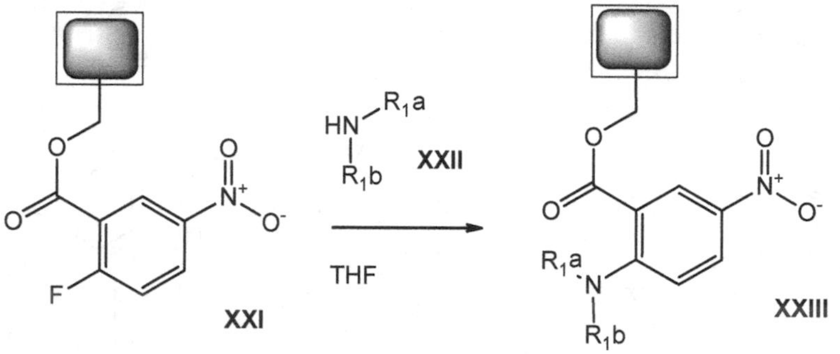

Scheme 2.10 Substitution of active fluorine by amine on solid support

The bags are labeled. Each vial is charged with a solution of a required amine dissolved in THF (5 equiv or 1.9 mmol per one portion of resin). The vials are agitated on a shaker for 18 h. The bags are washed successively with 2 × THF, DMF, 2 × MeOH, 2 × DCM, and 2 × hexane. As a result, fluorine in resin **XXI** is substituted with the corresponding dialkylamino group to form resin **XXIII**. This stage involves one important operation (capping). The case in point is that secondary amines partially destroy the ester bond which links the substituted benzoic acid to polymer. As a result, some part of hydroxymethyl groups of the resin get free and can function as a competitive center at the subsequent acylation stage. Since the acylating agents at stage 4 are derivatives of aromatic acids, the introduced acyl residues will persist on the resin until the last stage and contaminate the final product at the linker cleavage stage. Therefore, the hydroxymethyl groups should be protected by a "harmless" residue of a water-soluble acid, say, acetic. In practice, the resins are placed in a pyridine solution, and then, acetic anhydride is added. The quantities of reagents are estimated as follows: 5 equiv (1.9 mmol, 0.204 ml) of pyridine in abs. DCM per one bag and 5 equiv (1.9 mmol) of acetic anhydride. The vials are agitated in a shaker for 18 h at a room temperature. The resins are washed in succession with 2 × THF, DMF, 2 × MeOH, 2 × DCM, and 2 × hexane and vacuum-dried to obtain products **XXIII**.

Stage 3. *Reduction of nitro group*: Since we used the same reaction for all resins **XXIII**, the labeled bags are put in one reaction vessel (Fig. 2.4d), after which a 2 M solution of $SnCl_2 \times 2H_2O$ in DMF (50 ml per 1 bag) was added and then shaken at a room temperature for 48 h. The bags are successively washed with 2 × DMF, 2 × DMF/H$_2$O (1:1 v/v), 2 × H$_2$O, 2 × DMF, 2 × MeOH, 2 × DCM, 2 × MeOH, 2 × DCM, and 2 × hexane and vacuum-dried to obtain resins **XXIV** with a free amino group (Scheme 2.11).

Stage 4. *Acylation of amino group*: The operation is performed in two stages. To prevent incomplete reaction, conditions to ensure double acylation of the aniline amino group are chosen. At the following stage, mild removal of one of the amino groups is accomplished, whereas the second amino group remains intact.

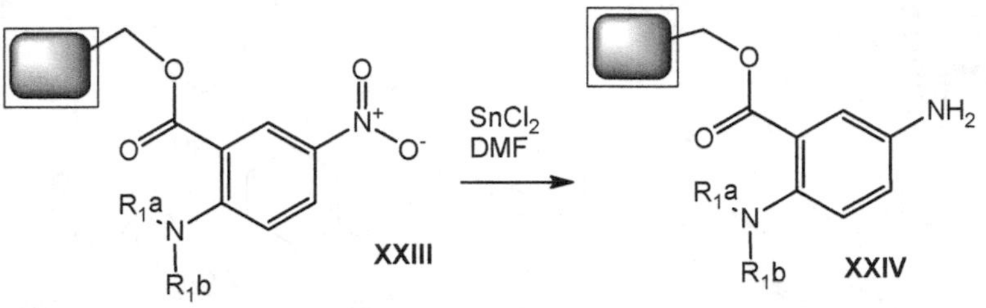

Scheme 2.11 Reduction of nitro group in nitrobenzene immobilized on solid support

Double acylation: According to the statements of work (which specify how many experiments to perform and what acylating agent to take), the bags with resins **XXIV** are again distributed over vials whose number is equal to the number of acyl chlorides **XXV** (Fig. 2.4c and Scheme 2.12). The bags are again labeled. To each vial, 5 equiv (1.9 mmol) of pyridine in abs. DCM and acyl chloride **XXV** (5 equiv or 1.9 mmol per one portion of resin) are added. The vials are agitated in a shaker at room temperature for 6–18 h. The bags are washed in a succession with 2 × DCM, 2 × DMF, 2 × MeOH, 2 × DCM, 2 × MeOH, 2 × DCM, and 2 × hexane and dried to obtain diacyl derivatives **XXVIa**.

Removal of one of the acyl groups: The bags with resins **XXVIa** are placed into a 25% solution of aqueous ammonia in DMF, agitated in a shaker at a room temperature for 18 h, and washed with 2 × DMF, 2 × MeOH, 2 × DCM, 2 × MeOH, 2 × DCM, and 2 × hexane to obtain monoacyl derivatives **XXVIb**.

Stage 5. *Removal of the Final Product from Support*: Each bag is placed into an individual vial, treated with 10% TFA/DCM, and left to stand at a room temperature for 2 h. The solution is decanted, and the solvent is removed at reduced pressure to obtain final products **XXVII**, as in Scheme 2.13.

Yield and purity of the obtained products: The isolable yields of products by 100 mg of the initial resin are, on average, 150 mg. General statistics of compounds' mass for a big library of compounds **XXVII** (∼ 840 samples) is shown on Fig. 2.5. The purity of final samples **XXVII**, determined by LC-MS (UV-254 and ELSD detectors), was >90% (Fig. 2.6).

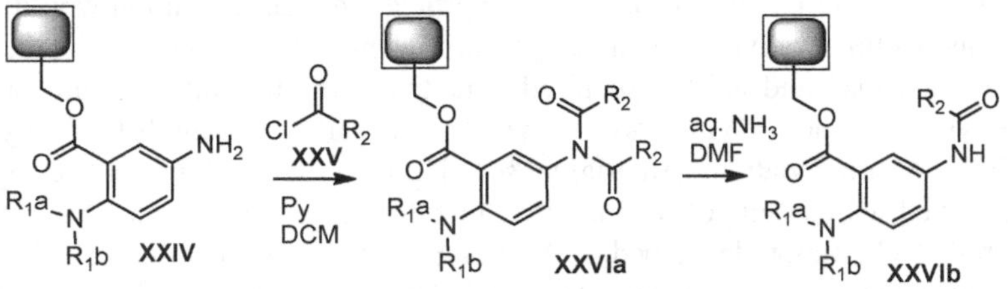

Scheme 2.12 Double acylation of the aniline and removal of one acyl group

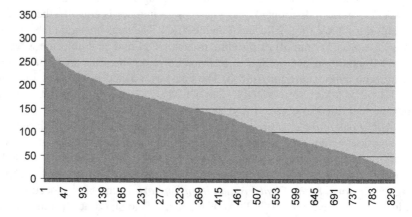

Scheme 2.13 Final removal of target compound from the tea bag

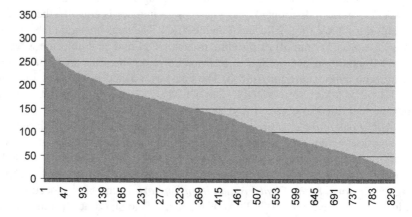

Fig. 2.5 Horizontal—no. of samples; vertical—amount obtained (see text)

2.3   Syntheses with Lantern Rods

The problem of polymer swelling (which complicated the transfer of resins from one vessel to another, or to a filter) is solved to success by using a filter tube "container" or a "tea bag." At the same time, one more inconvenience takes place: A fairly large volume of solvents is required to thoroughly wash out excess reagents from pores and internal volume of polymer particles. An ideal solution would be to use a non-swelling polymer such as polypropylene. Unfortunately, the desired functional groups (and in the required quantity) are difficult to graft on the polypropylene surface. (With polystyrene, this is readily accomplishable by, for example, chloromethylation.) An original solution is to use a polypropylene support surface-coated with a thin polystyrene film. Just this principle forms the basis of the lantern rods technology, developed by Mimotopes Pty Ltd (Australia) [6], Fig. 2.7.

Lantern rod is a plastic polypropylene cylinder coated with chemically modified polystyrene. To increase the total volume, an internal cavity and slots are made in the cylinder surface (thus making it similar to a Chinese lantern, as in Fig. 2.8a).

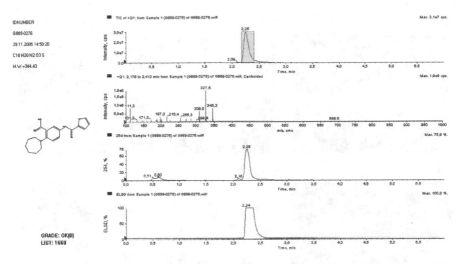

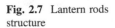

Fig. 2.6 Example of the purity of final sample **XXVII**

Fig. 2.7 Lantern rods structure

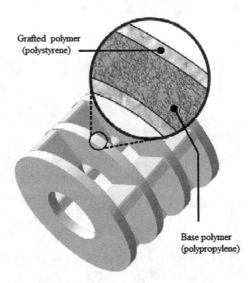

Grafted polymer (polystyrene)

Base polymer (polypropylene)

The dimensions of lanterns are chosen so that they fit the cavity of standard plastic billets (8 × 12). By immersing lantern into various reagents (Fig. 2.8b) and washing it every time with solvents (Fig. 2.8c), one can perform a multistage synthesis. As a rule, lanterns are used once and thrown away after the last stage ("cutting off" a complex final product from the support).

Lanterns facilitate the technique of multistage syntheses of large combinatorial libraries, and the quantities of substances, isolated at the last stages (5–50 mg), are

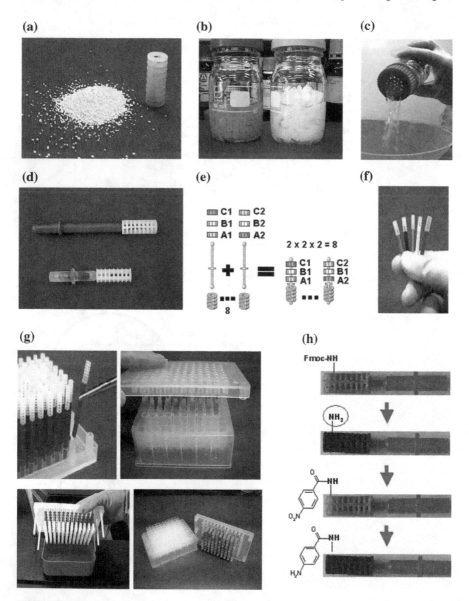

Fig. 2.8 Technique of the solid-phase synthesis on lanterns (see text)

more than sufficient for biological tests. The main problem of the lantern technique is logistic in nature, since one should learn how to record correctly the "fate" of each cylinder (not to disorder cylinders in the course of multistage syntheses). To this end, each lantern should be labeled in its specific way. Three approaches are known. The most reliable (and costly) is the so-called Irory technology, when each lantern is provided with a microchip (Fig. 2.8d), and a special electronic device is required to record and read off from microchips to which reactions each lantern should be introduced in (or has already been introduced in). A simpler approach is to fix lanterns in the cap of a billet with numbered holes and keep a record by lantern numbers. And, finally, one can fix colored plastic pins on lanterns and mark the reagents already used for synthesis with colored labels or rings (Fig. 2.8e–g). This approach is convenient in the case of small libraries.

2.3.1 · Choice of Model Reaction

A great number of examples of efficient usage of lantern in the synthesis of diverse libraries are reported [7–10]. Our attention was attracted by a simple synthetic sequence leading to N,N'-diarylureas [7] (Fig. 2.9a).

The key stages of this process are quite simple: A m-(or p-) nitrobenzoic acid residue is fixed on a lantern, the nitro group is reduced, and then, the aniline fragment is reacted with isocyanates. Attempting to correlate the structure of expected products with their potential biological activity, we noticed that close structural relatives of diarylureas, viz. diaryl- and triarylguanidines, are well-known medicines (Fig. 2.10). (Note that thiformine on Scheme 1.1 and its structural analogs in the libraries, too, contain the guanidine motif.)

It was not too difficult to associate the possible synthetic route to guanidines with the aniline intermediate in Fig. 2.9a: A thiourea residue can be obtained in a high yield by simply replacing isocyanates with isothiocyanates (Fig. 2.9b). This procedure is well-developed in a usual liquid-phase version [11]. Thioureas, in their turn, can be easily converted into guanidines by well-developed (again, in solutions) published procedures [12]. Just this sequence (probably, never realized in the solid-phase version) formed the basis of a task at the special student laboratory at the MSU.

Required reagents and equipment: Lanterns, 1.5-ml Eppendorf-type plastic tubes (50), and great number of serum vials are required. For removing the protective group from the resin, a 20% (v/v) solution of piperidine in DMF is prepared. The consumption of reagents is not very high, mostly 1–5 g. Triethylamine and isomeric *meta*- and *para*-nitrobenzoyl chlorides (the labels on the bottles are A and B) are required for the acylation stage. Reduction is performed using $SnCl_2 \times 2H_2O$. The isothiocyanates used are *meta*-chlorophenyl isothiocyanate [13] and furfuryl isothiocyanate [14], which were prepared by known procedures. To convert thioureas into guanidines, $NaIO_4$ and three amines are required: morpholine (α), m-toluidine (β), and benzylamine (γ). For the cleavage stage, we take trifluoroacetic acid; lanterns are washed with methanol and DCM.

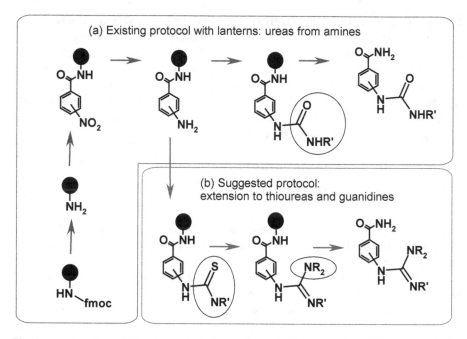

Fig. 2.9 **a** Procedure of diarylurea synthesis, recommended by the producer [7] and **b** synthetic protocol for preparing thioureas and guanidines, optimized as a training task for students

Timegadine
antiinflammatories

Phenodianisyl
local anesthetics

Aptiganel
neuroprotectives

Fig. 2.10 Drugs of the di(tri)arylguanidine series in the Merck Index

2.3.2 *Transformation Sequence and Methodical Notes*

In the work with lanterns, one of the main disadvantages of solid-phase synthesis, viz. difficult control of the degree of progress of multistage reactions, becomes the most evident. Spectral methods are unsuitable. (Actually, plastic pins are difficult to place into spectrometer ampoules or cells). At the same time, accumulation of incompletely reacted functional groups (at five or six planned stages) may dramatically affect the purity of the final product. A stage-to-stage purity control is possible: by "cutting off" intermediates from the support after every further stage.

This method is difficult for another reason: Only in the case of finite substance (with high molecular weight) that one can expect to discharge significant quantities of product, while low molecular weight intermediates are negligible.

Reaction progress is optimally (and what is more, visually) controlled as follows. Let us note (see Fig. 2.9) that at the first four stages, the amino group is either present on the resin surface (the first time after protection removal and the second time after NO_2 reduction) or absent (on the initial lanterne, in the nitrobenzamide fragment, and in aryl thiourea). Such an alternating appearance and disappearance of the amino group can be detected by means of the Kaiser test (see Chap. 1). It was found (see Fig. 2.8h) that this test is quite effective; i.e., it makes sense to use a control lantern at each stage (develop it with ninhydrin and then throw away). In practice, lanterns are sprayed with a ninhydrin solution and heat for a short time with a hot air (heat gut or hair drier). In the case of diaryl thioureas, control removal of an intermediate from the support and assessment of its purity by NMR were performed. The purity of final guanidines was controlled by TLC, and their composition was determined by mass spectrometry (direct inlet).

The overall sequence and logistics of transformations on 7 lanterns (plus 4 control lanterns) is shown in Scheme 2.14. Let us consider the stages in more detail:

(1)   Protection removal: Lanterns containing the Rink resin (R–CH$_2$–NH–Fmoc) were treated with a piperidine solution to remove the protective group. The ninhydrin test on a control lantern is positive.

(2)   Acylation of the free amino group of lantern **LX** (see the denotation in Scheme 2.14) with *para-* and *meta*-nitrobenzoyl chlorides **A** and **B** gives resins **LXA** and **LXB**. The ninhydrin test on a control lantern is negative.

(3)   Reduction of the nitroaryl group to aniline with SnCl$_2$. (The codes of the resulting substances are unchanged.) The ninhydrin test on a control lantern is positive.

(4)   Formation of diaryl thioureas **LXAI**, **LXAII**, and **LXBII** by the reaction of the amino group with isothiocyanates **I** and **II**. The ninhydrin test on a control lantern is negative.

(5a) Cutting-off of diaryl thioureas **XAI**, **XAII**, and **XBII** with trifluoroacetic acid (TFA), determination of the yield (about 10 mg), and measurement of the NMR spectrum.

(5b) Treatment of the lanterns from experiment (4) with amines α, β, and γ. Preparation of guanidines **LXAI1**, **LXAI2**, **LXAI3**, and **LXAII1**.

(6)   Cutting-off of guanidines **XAIα**, **XAIβ**, **XAIγ**, and **XAIIα** with TFA (about 10 mg) and mass spectral analysis of the isolated products.

Scheme 2.14 shows consecutive steps which were undertaken to *optimize* the task. Theoretically, $2 \times 2 \times 3 = 12$ substances would be expected. The absence of certain combinations is explained by the fact that что *p*-nitrobenzoyl chloride **B** forms an unsatisfactorily unclear intermediate, which was found out at an early purity control stage (analysis of thioureas). Thiourea **XAII** (from furfuryl isothiocyanate **II**) gave a satisfactory ^{1}H NMR spectrum (Fig. 2.11a), but analysis of the guanidine obtained

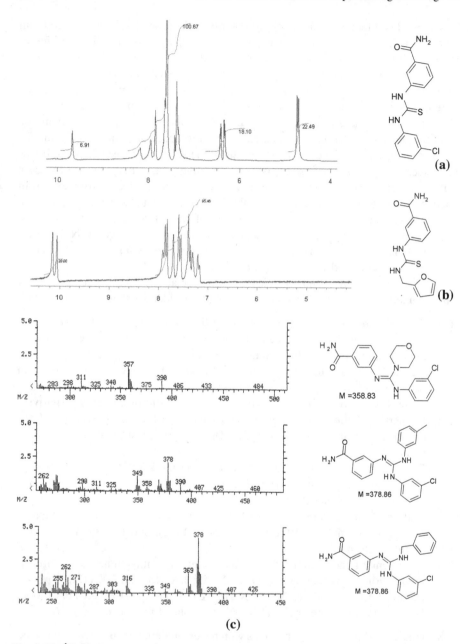

Fig. 2.11 ¹H NMR spectra of **a** thiourea **XAII** and **b** thiourea **XAI**. **c** Mass spectra of final guanidines **XAIα**, **XAIβ**, and **XAIγ**

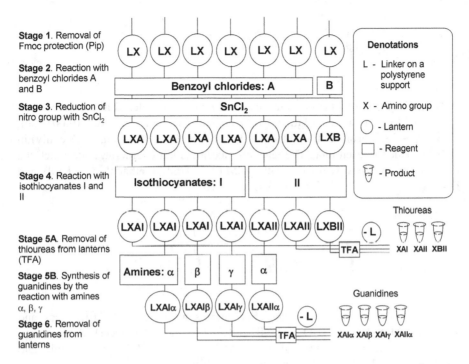

Stage 1. Removal of Fmoc protection (Pip)

Stage 2. Reaction with benzoyl chlorides A and B

Stage 3. Reduction of nitro group with SnCl₂

Stage 4. Reaction with isothiocyanates I and II

Stage 5A. Removal of thioureas from lanterns (TFA)

Stage 5B. Synthesis of guanidines by the reaction with amines α, β, γ

Stage 6. Removal of guanidines from lanterns

Scheme 2.14 Work with lanterns: overall sequence and logistics of transformations (see text)

from this thiourea revealed a great number of admixtures (probably, formed by side reactions involving the furan ring). In this regard, we suggested to students for practical implementation only a part of Scheme 2.14, specifically, to use one benzoyl chloride (**A**), one isothiocyanate (**I**), and three amines, to obtain one thiourea (**XAI**) and three guanidines (**XAIα**, **XAIβ**, and **XAIγ**).

2.3.3 Preparation of N,N′,N″-Trisubstituted Guanidines on Lanterns

Before lesson, the teacher demonstrates a negative Kaiser test for a standard lantern. Each student obtains 8 lanterns.

Stage 1. *Removal of Fmoc protection*: Eight lanterns are placed in 1.5-ml plastic tubes (Eppendorf-type), and 1 ml of 20% piperidine in DMF is added to each tube. After 1 h, the sticks are successively washed with DMF (3 × 1 ml), methanol (3 × 1 ml), and DCM (3 × 1 ml) and transferred into new tubes. The control lantern shows a positive Kaiser test.

Stage 2. *Acylation of amino group*: Seven lanterns are placed into plastic tubes, and 1 ml of a 1 M solution of *meta*-nitrobenzoyl chloride (0.001 mol, 0.185 g, ~ 28 equiv) and triethylamine (0.001 mol, 0.1 g, 0.14 ml, ~ 28 equiv) in DCM is added to each tube. The reaction mixture is left overnight, after which the lanterns are successively washed with DCM (3 × 1 ml), methanol (3 × 1 ml), and DMF (3 × 1 ml). The Kaiser test was negative.

Stage 3. *Reduction of nitro group*: Six lanterns are placed into plastic tubes, and 1 ml of 1 M $SnCl_2 \times 2H_2O$ in DMF (i.e., 0.001 mol of reducer). After a day, lanterns are successively washed with DMF (3 × 1 ml), DCM (3 × 1 ml), and methanol (3 × 1 ml). The Kaiser test was positive.

Stage 4. *Coupling with isothiocyanate*: Five lanterns are placed into plastic tubes, and 1 ml of a 1 M solution of *meta*-chlorophenyl isothiocyanate (0.001 mol, 0.169 g, 0.14 ml) in methanol is added to each tube. They are left overnight and then successively washed with methanol (3 × 1 ml), DCM (3 × 1 ml), and DMF (3 × 1 ml). The Kaiser test is negative.

Stage 5A. *Cleavage of amide and separation of final thiourea XAI*: One of the lanterns is placed into a plastic tube, 1 ml of a 20% (v/v) solution of TFA in DCM is added, and the mixture is left to stand for 1 h. The solution is transferred to a serum vial, and the lantern is additionally washed with DCM (3 × 1 ml). The combined fractions is first concentrated in air, transferred in a tube, and further concentrated in a vacuum drying oven. *N*-(*m*-Chlorophenyl)-*N′*-(*m*-phenylcarbamide) thiourea (**XAI**), 10 mg, R_f = 0.7 (Silufol, CHCl₃/MeOH = 8/2), is obtained. According to TLC, the product contains no admixtures. The ¹H NMR spectrum (Fig. 2.11a) contains no extra signals and completely fits the structure. The same is true for furane isothiocyanate (Fig. 2.11b).

Stage 5B. *Guanidination*: The three remaining lanterns are placed into plastic tubes labeled 1, 2, and 3, and 1 ml of a freshly prepared solution of NaIO₄ (0.1 g, 0.00046 mol) and then a solution of corresponding amine α − γ (0.0007 mol) in DMF are added to each tube. The reaction mixtures are heated for 6 h at 60 °C and left overnight at room temperature. Precipitate formation is observed in all the cases. The lanterns are washed with water (5 × 1 ml), DMF (3 × 1 ml), methanol (3 × 1 ml), and DCM (3 × 1 ml).

Stage 6. *Cleavage of amide and separation of the final guanidine*: The tubes with lanterns are charged with 1 ml of 20% TFA in DCM and left to stand for 1 h. The solutions are transferred to serum vials with numbers, and the lanterns are additionally washed with DCM (3 × 1 ml). The combined fractions are concentrated first in the air, transferred into new weighed tubes, and further concentrated in a vacuum drying oven to isolate 8–12 mg of the final guanidine. All the three products have close R_f values (∼0.4) on Silufol (CHCl₃/MeOH = 8/2).

By TLC, a minor admixture (5–10%) with R_f = 0.8 was detected. No further purification was performed. The mass spectra show molecular ion peaks with expected *m/z* (Fig. 2.11b).

2.4   Conclusions

Three examples of tools facilitating manipulations with polymer supports were considered. In the first example, we used a container for resins and the Bill-Board kit (reaction vials are combined with filters) and successfully obtained the library of glycines by consecutive C-alkylation and N-acylation. In the second example, we used the tea-bag technique (resin in a plastic bag) and we were able to obtain quite big library of the amino amides in 5 steps. In the third example, we considered the lanterns, specially modified polypropylene pins, and obtained in 6 stages the libraries of thioureas and guanidines. All the three methodologies are comparable, and all of them represent versions of training tasks.

References

1. O'Donnell MJ, Zhou C, Scott WL (1996) Solid-phase unnatural peptide synthesis (UPS). J Am Chem Soc 118:6070–6071. doi:10.1021/ja9601245
2. (a) Scott WL, Zhou C, Fang Z, O'Donnell MJ (1997) The solid phase synthesis of α,α-disubstituted unnatural amino acids and peptides (di-UPS). Tetrahedron Lett 38: 3695–3698. doi:10.1016/S0040-4039(97)00715-6. (b) Alsina J, Scott WL, O'Donnell MJ (2005) Solid-phase synthesis of α-substituted proline hydantoins and analogs. Tetrahedron Lett 46: 3131–3135. doi:10.1016/j.tetlet.2005.01.163
3. Scott WL, O'Donnel MJ (2009) Distributed drug discovery, part 1: linking academics and combinatorial chemistry to find drugs for developing world diseases. J Comb Chem 11:3–13. doi:10.1021/cc800183m
4. Scott WL, Alsina J, Audu CO, Dage JL, Babaev EV, Cook L, Dage JL, Goodwin LA, Martynow JG, Matosiuk D, Royo M, Smith JG, Strong AT, Wickizer K, Woerly EM, Zhou Z, O'Donnell MJ (2009) Distributed drug discovery, part 2: global rehearsal of alkylating agents for the synthesis of resin-bound unnatural amino acids and virtual D3 catalog construction. J Comb Chem 11:14–33. doi:10.1021/cc800184v
5. Houghten RA (1985) General method for the rapid solid-phase synthesis of large numbers of peptides: specificity of antigen-antibody interaction at the level of individual amino acids. Proc Nat Acad Sci 82(15):5131–5135. doi:10.1073/pnas.82.15.5131
6. www.mimotopes.com
7. www.mimotopes.com/files/editor_upload/File/CombinatorialChemistry/SynPhase_Publications.pdf
8. Bray AM, Valerio RM, Dipasquale AJ, Greig J, Maeji NJ (1995) Multiple synthesis by the Multipin method as a methodological tool. J Pept Sci 1(1):80–87. doi:10.1002/psc.310010110

9. Gerritz SW, Norman MH, Barger LA, Berman J, Bigham EC, Bishop MJ, Drewry DH, Garrison DT, Heyer D, Hodson SJ, Kakel JA, Linn JA, Marron BE, Nanthakumar SS, Navas FJ (2003) High-throughput manual parallel synthesis using SynPhase crowns and lanterns. J Comb Chem 5(2):110–117. doi:10.1021/cc020022c

10. Parsons JG, Sheehan CS, Wu Z, James IW, Bray AM (2003) A review of solid-phase organic synthesis on SynPhase lanterns and SynPhase crowns. Methods Enzymol 369:39–74. doi:10.1016/S0076-6879(03)69003-8

11. Mozolis VV, Iokubaitite SP (1973) Preparation of N-substituted thiourea. Usp Khim 42(7):1310–1324. doi:10.1070/RC1973v042n07ABEH002677

12. Ramadas K, Janarthanan N, Pritha R (1997) A short and concise synthesis of guanidines. Synlett 9:1053–1054. doi:10.1055/s-1997-1535

13. Radl S (1992) Preparation of some pyrazole derivatives by extrusion of elemental sulfur from 1,3,4-thiadiazines. Coll Czech Chem Commun 57(3):656–659. doi:10.1135/cccc19920656

14. Spurlock LA, Fayter RG (1969) Nature of the carbonium ion. II. Furfuryl cation from a thiocyanate isomerization. J Org Chem 34(12):4035–4039. doi:10.1021/jo01264a061

Chapter 3
Secrets of Parallel Liquid-Phase Synthesis

In the previous chapters, we considered the possibilities of solid-phase organic synthesis (SPOS) of libraries of chemical compounds, and therewith observed that some of these methods can be used as training tasks at students' laboratories. All manipulations in SPOS are in fact limited to two multiply repeated operations: addition of an outside reagent (reaction as such) and its removal (filtering and washing). The technology may slightly vary, but a common feature of all variants is that a reactor is combined with a filter for purification. In this sense, the difference between a tube with a filter (Bill-Board) and a porous plastic container ("tea bag") is not too big. A lantern combines both of these principles, and filtration comes down to routine washing of a macroscopic object.

The advantages of SPOS over usual liquid-phase organic synthesis (LPOS) are undeniable. These are as follows: (1) the access to large combinatorial libraries with moderate expenses, (2) possibility of automation of the process, (3) higher yields of the target product due to a large excess of reagents, and (4) facility of purification, because the excess reagents, by-products, and salts are readily removed at the washing stage. The method is ideally suited in cases when one has, at different stages, to combine or isolate intermediate compounds, involving them in complicated reaction sequences. The SPOS technology allows successful labeling of multiply changed objects; this relates to so-called tagging technologies.

If conditions are thoroughly selected and optimized (this may take 1–3 months; Fig. 3.1), SPOS can provide more than 500 compounds per month. However, the SPOS methodology has certain limitations. First, only a relatively small fraction of reactions is adapted to occur on supports. It is hard to imagine that many traditional reactions (sulfonation, nitration, and even bromination) with a molecule grafted on a polystyrene support (It should also not be forgotten that linkers are sensitive to acids and alkalis.) Second, introduction of "new" reactions into SPOS takes a long time, since the entire synthetic sequence (including separation of support and

© The Author(s) 2017
E.V. Babaev, *Incorporation of Heterocycles into Combinatorial Chemistry*,
SpringerBriefs in Molecular Science, DOI 10.1007/978-3-319-50015-7_3

Fig. 3.1 Approximate productivity of different methods of multistep synthesis (adapted from [1])

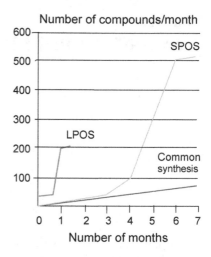

product) should be perfectly worked out. Third, solid reagents and insoluble catalysts cannot be applied in SPOS. Fourth, SPOS is accomplished almost "blindly," since intermediate products are difficult to identify. Finally, many popular supports and linkers of SPOS are expensive.

Thus, it is most frequently reasonable and at times unavoidable to accomplish parallel reaction in a usual LPOS variant. (For a scientist working with liquids by hands who physically observes experiments, the commonly accepted term is wet chemistry). Therewith, it takes less time to optimize LPOS compared to SPOS (Fig. 3.1). In practice, both methodologies complement each other. Development of parallel LPOS technologies already has its own history (see the review [1]), and some most illustrative details are discussed in the present review.

3.1   Technology of Parallel LPOS

The essential attributes of classical LPOS are multinecked flasks with reflux condensers, dropping funnels, and thermometers, which are usually heated on magnetic stirrers. Obviously, to perform tens and hundreds reactions in parallel, an alternative to this conventional laboratory equipment is required. In view of our long-term experience in training students (laboratory course on combinatorial chemistry at the MSU), at this point, we would like to consider both the most modern expensive equipment and routine facilities (the classical principle of "sealing wax and rope") for a low-budget laboratory.

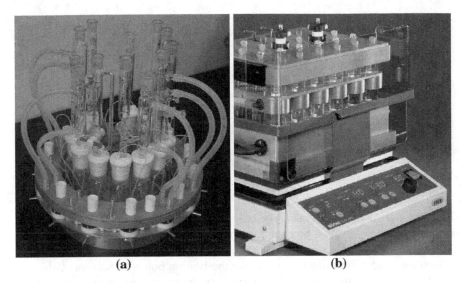

(a) **(b)**

Fig. 3.2 Parallel refluxing of a great number of reaction vessels in **a** CombiSyn and **b** SynCore

3.1.1 Parallel Refluxing

It is observed that a great number of patented solutions as to how to combine several reaction vessels into a single construction are available. A smart solution is our tested apparatus CombiSyn (Fig. 3.2a). The apparatus was designed by M. Baru (Pushchino, Moscow Region, Russia) and was previously distributed via Chemical Diversity. As we have seen, ordinary reflux condensers have a common circulation system, which allows a compact arrangement of ordinary reaction vessels. The apparatus functions by the carousel principle and up to 13 reaction vessels can be refluxed and stirred by the means of one single magnetic stirrer.

The second model we tested is a Buchi parallel synthesis apparatus SynCore (Fig. 3.2b). This model is envisioned to accommodate 24 reaction vessels. Heating and stirring are performed by means of a temperature-controlled shaker. The SynCore apparatus is equipped with a single reflux condenser (a hollow metallic module with 24 slots), which is mounted on cylindrical reactors. All vessels are covered on the top with a common metallic cap, through which one can add reagents or force an inert gas. The main "secret" of the cap, allowing evaporation of all the 24 vessels simultaneously, will be discussed below.

3.1.2 Parallel Boiling Without Condensers

High-temperature boiling of solutions has as long as two centuries been performed in the simplest autoclaves, viz. sealed tubes, this idea is still widely exploited in

Fig. 3.3 Vials for parallel synthesis, a substitute of a reflux condenser

combinatorial chemistry till now. The only thing is that the reaction vessel here is a thermal-glass tube or plain-bottomed vial with a tight screw cap (Fig. 3.3).

As a rule, the cap for a heated vial has a Teflon rubber seal. Such construction is quite strong and allows overheating above the boiling points of solvents (which is actually equivalent to using sealed ampoules). After cooling, excess pressure is readily released by piercing the seals. In our experiments, vials with acetonitrile solutions (mp 82 °C) tolerated many-hour heating at 100 °C. A few vials can be stirred with a usual magnetic stirrer without heating (Fig. 3.4a). For a large number of vials, one can make a simple support, fix it on a shaker, and bottom-heat vials by means of an electric heating coil attached to a standard laboratory transformer (Fig. 3.4b, after D. Al'bov). The safest approach is to heat hermetically sealed vials in the holes of temperature-controlled block heaters whose design (and well shapes) may vary. We successfully tried a J-Kem 310 block heater of a 10 × 10 format (Fig. 3.4c) with a timer and electronic controller of temperature and shaking intensity.

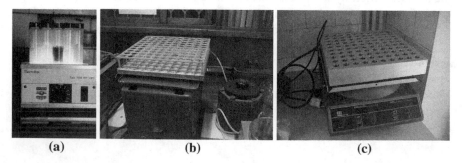

(a)   **(b)**   **(c)**

Fig. 3.4 Techniques for parallel heating of vials (see text)

3.1.3 Synthesis in Microwave Ovens

Reactions in microwave (MW) ovens are shorter and thus meet the main requirement of combinatorial chemistry, namely acceleration of synthesis. There are two ways to perform such reactions. The first is to perform a great number of reactions simultaneously. A rotating support (carousel) loaded with a great number of reaction vessels is placed into an MW oven (Fig. 3.5a–c). The magnetic stirrer built-in into the bottom of the apparatus provides a uniform stirring in each vial. In the second technology, the MW oven contains only one port for one vessel (Fig. 3.5d), and reactions are performed sequentially by means of a robotic module: The vials are changed in a computer-controlled sequence. Technical solutions are constantly improved: in the considered CEM models, the temperature in the reference vessel is measured with an optical fiber sensor. Automatic shutdown of heating is also applied when the pressure in vessels gets rising.

3.1.4 Syntheses in Flow Microreactors

Among the most recent technical innovations for synthesis acceleration, we would like to mention the technologies which are based on flow microreactors. One hardly imagines a usual reaction vessel without a powerful mechanical or magnetic stirrer,

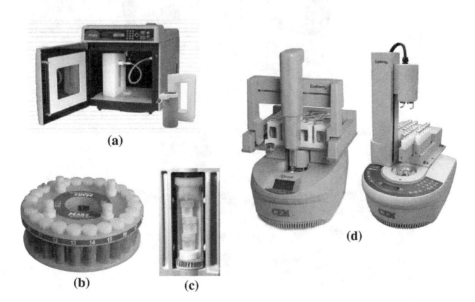

Fig. 3.5 Microwave ovens for combinatorial synthesis: **a** MW oven with a carousel for **b** glass and **c** plastic vessels in the MARS CEM model; **d** robotized attachments for fast sequential synthesis in the CEM explorer model

as effective stirring is an indispensable condition for successful reaction. In the examples above, the same role was played by agitation in a shaker. In its essence, such stirring is turbulence, and its speed has a certain limit. This limit can be overcome by the laminar diffusion. To induce this phenomenon, one has to make the volume where reagents meet each other close to zero (by performing reaction in a very thin capillary, Fig. 3.6a), and the rate of this meeting, to infinity (by applying a very high pressure to inject reagents into this volume).

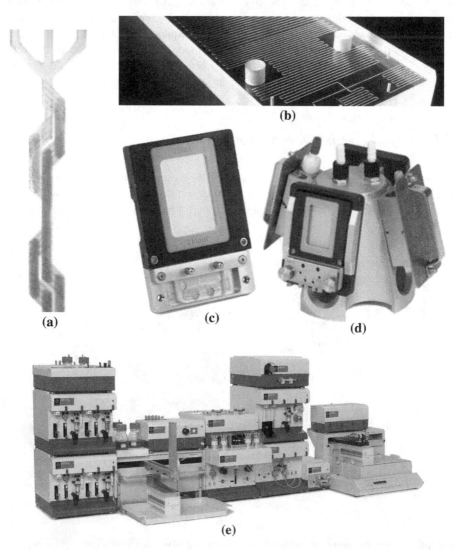

Fig. 3.6 Flow microreactors for fast synthesis: **a** capillaries; **b**, **c** quartz or plastic plates; **d** block of several plates with capillaries; and **e** general view of a Syrius Africa microreactor

Just this principle is realized in flow microreactors whose "heart" is formed by powerful pumps and "arteries," by extremely thin capillaries engraved (or chemically etched) inside quartz or plastic plates (Fig. 3.6b, c). Now, the highest degree of reagent mixing has time to occur in the first millimeters of the capillary, taking fractions of a second. For the reaction to come to completion, the capillary should be long enough (a meter and longer). The plate with a capillary can be likened to a sealed vial (by "locking" the outgoing liquid by counterpressure at the outlet). Thus, the flowing solvent can be overheated above its boiling point, thereby still more accelerating the process.

In up-to-date microreactors, the reagent feeding and product collection system is robotized, and, therewith, for parallel synthesis several plates with capillaries are used (Fig. 3.6d). As a result, the synthesis is made fully automated and feasible for preparing both small libraries (on the microscale) and tens and hundreds grams of a target product. Such "factories" (really substituting the whole plant floors) can be easily accommodated on a desktop. Figure 3.6e shows a fully robotized system Africa (Syrris) realizing this technology, which was purchased for our research. A great number of traditional reactions have already been optimized (see the specialized journal Lab on Chip). Now these reaction times are unimaginably short, and the yields are sharply increased.

3.2 Separation of the Reaction Mixtures

Any parallel synthesis is inevitably followed by the stage of treatment of the reaction mixture, not infrequently quite a laborious process. The heterogeneous reactions with precipitate formation require filtration, and homogeneous reactions generally require extraction (either of the reaction mixtures in themselves or after their dilution with water). In phase transfer reactions, too, phase separation is required.

3.2.1 Parallel Extraction

It is not very difficult to add water (or another solvent) simultaneously into several vessels. Vigorous parallel agitation of emulsions in a shaker, too, is not quite a complicated problem. More difficult is to separate phases in parallel not resorting to classical separatory funnels. Several approaches are possible. One can act in a slightly unusual way and not to go beyond filtration or decantation. The aqueous layer in the vial can be frozen out in a refrigerator and easily separated from the organic layer, since water freezes easier than other solvents used for extraction. Another way is to pierce a flexible plastic vial cap with two syringe needles (Fig. 3.7a).

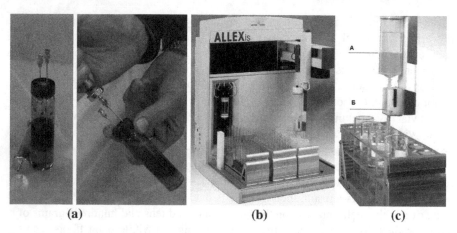

Fig. 3.7 Parallel extraction: **a** with a syringe; and **b**, **c** workstation for parallel synthesis

One needle is used to withdraw the required phase with a syringe, and the other needle (not immersed into the liquid) serves for pressure equalization. Separation can also be accomplished using high-cost automated workstations for parallel synthesis (Fig. 3.7b, c). A two-phase system is passed through a special sensor (B) sensitive to physical parameters at the phase interface (A), and the liquid flow is stopped in a due moment.

Emulsions can also be separated by means of a so-called molecular filtering (realized in the Africa auto synthesizer as a FLLEX attachment, Fig. 3.8a). Water is injected into the organic liquid flow (or vice versa), the resulting aqueous organic emulsion is forced under pressure through a membrane, and then again it is separated into the components (Fig. 3.8b). The plastic membrane is quite small (Fig. 3.8c), and its pore size is so that only small water molecules can pass through (Fig. 3.8d).

3.2.2 Parallel Filtration

If precipitates are present in the reaction vials, it would be desirable to filter them in parallel, not transferring suspensions into a filter and using a single pump. Reaction vessels combined with filters are in general use in solid-phase and liquid-phase syntheses. If a reaction occurs at room temperature and results in precipitate formation, it can be performed in a 96-well plastic plate (Fig. 3.9a–c). Therewith, the billet bottom should be porous (by the principle "vessel = filter"). During the reaction, the billet is hermetized on the top and bottom with flexible seal caps. After the reaction, the caps are removed (Fig. 3.9a) and apply special facilities to vacuum filter 96 precipitates (Fig. 3.9b, c) and to collect 96 filtrates.

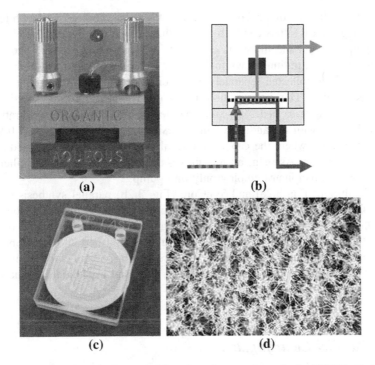

Fig. 3.8 Separation of a two-phase emulsion through a membrane in the FLLEX attachment in the Africa system. For denotations, see text

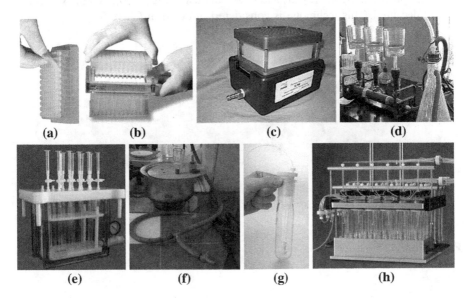

Fig. 3.9 Systems for parallel filtration. For denotations, see text

For parallel filtration on a laboratory scale, "filtration lines" with a few filters are produced (Fig. 3.9d). Filters can be fixed in different ways (for example, with usual rubber stoppers), and the filtrate is collected in a common receiver. The simple system for parallel filtration shown in Fig. 3.9e deserves special mentioning. Here, each filtrate can be collected separately, by placing the support with receivers into a vacuum desiccator. This construction is also feasible for drying solutions by passing through a drier bed. The simplest device for simultaneous filtration of several solutions, with a standard vacuum desiccator and standard glass filters is shown in Fig. 3.9f, where the cap with several slots for filters is equipped with a single vacuum valve. As seen, all the above-mentioned techniques for filtering a great number of precipitates employ only one pump.

The CombiSyn (Fig. 3.2a) and SynCore (Fig. 3.2b) parallel synthesis systems realize a slightly different principle. In the first case (Fig. 3.9g), a plastic tube with a filter at the end is immersed into the vessel, and the solvent is pumped out in a vacuum. In the second case (Fig. 3.9h), filter tubes are pushed into each tube up to the very bottom, a gas pressure is applied, as a result, the solvent is forced through a system of prelabeled tubings into 24 receivers. We used just this technique in our training task (see Chaps. 4 and 5).

3.2.3   Parallel Centrifugation

Parallel centrifuging is widely used for precipitate separation. To this end, attachments for great number of vessels and billets are used (Fig. 3.10), and the solvent is simply decanted. This technique is considered in detail in Chap. 5.

Fig. 3.10   Attachments for parallel centrifugation

3.2.4 Evaporation of Solutions

Simultaneous evaporation of several solutions is likely to present the most serious problem of parallel LPOS. In cases where special equipment is unavailable, there are three possible solutions: (1) sequential evaporation of a great number of samples on rotary evaporators; (2) long-term evaporation of open reaction vessels at ambient temperature and atmospheric pressure (for example, under a hood for light solvents); and (3) use of a vacuum drying oven with a careful vacuum control to prevent solvent bumping.

The listed methods are unsuitable if one needs to evaporate large series of solutions in a high-boiling solvent (for example, DMSO which is frequently used in biological tests). Analogous situation is when one evaporates fairly large volumes (10–100 ml) of extracting solvents. In such cases, one needs either a vacuum centrifuge or a system for parallel evaporation in a low vacuum. In a vacuum centrifuge (Fig. 3.11a), we can place individual vessels, billets with solutions, or plate holders with tubes. Bumping and cross-contamination are avoided due to a centrifugal force (sometimes billets are covered with a "net"). The evaporation rate can be increased by freezing out the distillate in the receiver. The SynCore apparatus (Figs. 3.2b and 3.11b) combines several techniques to prevent bumping and for preventing bumping and cross-contamination of solutions in neighboring vessels. Vacuum is set and maintained with an electronic controller. A hermetically sealed cap directs vapors through a sophisticated labyrinth of slots in the upper part of the cap, and therefore, the splashed liquid returns back into the same vessel.

3.2.5 Parallel Purification

Purification of a great number of reaction mixtures is an unavoidable problem of LPOS. To this end, the laboratories of major pharmaceutical companies make use

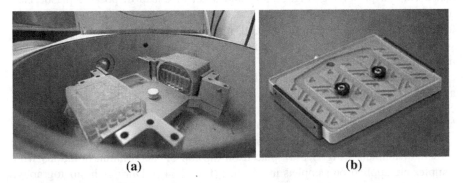

(a)   (b)

Fig. 3.11 Parallel evaporation by means of **a** vacuum centrifuge or **b** special hollow attachment to the Syncore apparatus, connected to a vacuum pump

of preparative liquid chromatographs with a programmed change of receiver at changing concentration of a substance passing through the column. Efficiency is provided due to full-time operation of several instruments. Unfortunately, the high cost of such equipment restricts its use at educational institutions. At the same time, there are a number of efficient low-cost techniques for parallel purification.

Solid-phase Scavengers considerably facilitate purification of single-type reaction mixtures. The scavengers are polymer supports with an active functional group (sulfonic acid, tertiary amine, etc.) on the surface. If some reaction components (for example, amine, base, or acid) were taken in excess and are present in the final reaction mixture, they are reliably removed by scavengers. Technical details of operation with a simple scavenger are discussed in [2].

Parallel Chromatography can be performed in different ways (Fig. 3.12). Low-cost billet filters (4 × 8 или 8 × 12) which are convenient to use not only for parallel filtration (Fig. 3.9a–f), but also for parallel chromatography (Fig. 3.12a). If chromatographic columns of a larger volume are needed, then glass column with solder-in filters are taken and compactly fixed on supports (Fig. 3.12b). Features of these technologies are discussed in Chap. 4 on a concrete example of reductive amination. Chromatography in plastic or glass columns can be accelerated by either evacuation of the receiver or supplying an inert gas from a balloon onto the top of the column (Fig. 3.12b). In both cases, there is a threat to "over dry" the column and disturb uniformity of the separation process. An interesting solution of this problem can be provided by a simple system for pumping of eluent through several columns simultaneously (Fig. 3.12c). One liquid supply pump and a simple system of distribution tubing for 5 columns are used. An automatic sample collector can be replaced by a convenient manual device (Fig. 3.12d).

Economical Flash Chromatography at rigid purity requirements to library compounds, consecutive chromatographic purification of every individual sample, rather than parallel chromatography is required. How can we drive this process? We tried a SepaCore economical and efficient flash chromatography system. The system comprises a portable preparative chromatograph (Fig. 3.13a) with an isocratic pump, UV detector, manual sample collector, mini recorder, apparatus for super-dense column packaging (Fig. 3.13b), and a set of cheap disposable plastic cartridges (Fig. 3.13c). Efficient column packaging is reached due to that a disposable tube column is evacuated from one end, immersed into a bed of silica or another sorbent, fluidized with nitrogen, and then tightly sealed with porous inserts by means of the device in Fig. 3.13b. To transform a plastic tube into an efficient column takes a few minutes. Such technique is readily applied for fast separation of a series of mixtures. In certain cases, the sorbent in a column is inexpedient to change (especially, if reversed-phase or costly chiral sorbents are used), while multiple successive runs pose a risk of damage.

The best solution in such situation is to use plastic disposable containers with sorbents (samplets) (Fig. 3.14a), placed on the top of the column. Solutions with samples are applied on samplets in advance (Fig. 3.14b). During chromatography, tarry products are retained on the insert, and after the separation is complete, the insert is replaced (Fig. 3.14c).

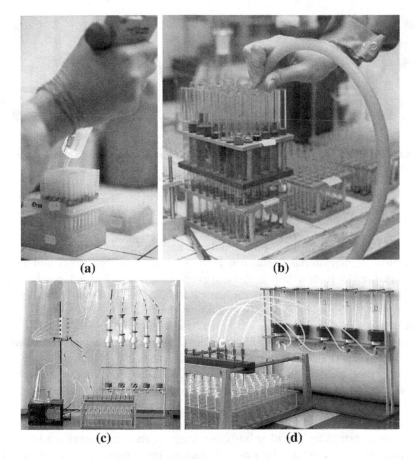

(a)   (b)

(c)   (d)

Fig. 3.12 Systems for parallel chromatography. For denotations, see text

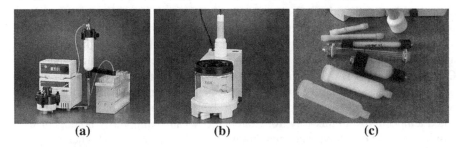

(a)   (b)   (c)

Fig. 3.13 Use of disposable columns in the SepaCore system. For denotations, see text

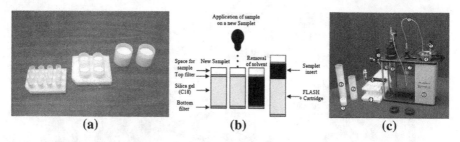

Fig. 3.14 Disposable *Samplet* cartridges for preparative flash chromatography. For denotations, see text

3.3 Conclusions

We would like to mention several trends in the development in techniques for liquid-phase combinatorial chemistry. First, equipment compactness and miniaturization (transfer from a flask to vial and even to capillary). Second, combining in a common space (vessel = filter = separatory funnel = column) traditionally incompatible processes (synthesis—workup—purification), which saves time. Third, sharp increase of the cost of the most efficient equipment (which is compensated for by the low cost of consumables). Therewith, glass is substituted by disposable plastic (billets, syringes, cartridges, samplets, etc.). Finally, the very idea of *parallel* processes is substituted by the idea of *sequential* processes which are, however, so fast (performed in an automated mode) that the total time consumption is reduced.

The LPOS techniques are constantly fed by new ideas from SPOS. One of the tendencies is extending use of solid-phase reagents and scavengers readily separable from reaction mixtures by simple filtration. Thus, solution pHs can now be controlled by acids and based immobilized on a support (i.e., not contaminating the solution with new mineral components), and oxidation can be performed by means of resin-supported oxidants (permanganates, periodates, etc.). Further trend is to perform reactions in a homogeneous medium and transfer them, when necessary, to a heterogeneous phase. To this end, one can use, for example, polyethylene glycol, an excellent solvent which can be precipitated, when needed, by adding methanol. One more example is perfluorinated solvents which do not mix with conventional organic solvents. It was found that heterogeneous emulsions on the basis of perfluorinated components magically transform into homogeneous at slightly elevated temperatures. In this way, for example, one can "temporarily homogenize" a required reagent or catalyst and then remove it (by imparting to it a higher affinity to the perfluorinated phase). As seen, the scope of traditional liquid-phase synthesis turns to be inexhaustible and is being more and more enriched with new scientific and technical solutions and inventions.

References

1. Miertus S, Fassina G (eds) (1999) Combinatorial chemistry and technology. Principles, methods and applications. Marcel Dekker, New York
2. Marsh A, Carlisle JC, Smith SC (2001) High-loading scavenger resins for combinatorial chemistry. Tetrahedron Lett 42:493. doi:10.1016/S0040-4039(00)01999-7

Chapter 4
Most Combinatorial LPOS Reaction: Reductive Amination with a Scavenger

4.1 Reductive Amination

Reactions of liquid-phase parallel synthesis are subject to quite stringent requirements by quite different criteria. Collections of starting reagents should be cheap
and diverse while the reaction conditions should be simple and mild, and the
products should be obtained in high yields and without complicated purification.
For biological tests, the final molecule should contain a predetermined "pharmacophore" fragment. Many reactions fail to meet such criteria. For example, certain
C–C bond formation reactions can be considered unsuitable. (Diels–Alder reactions
and even condensations produce a lot of admixtures; cross-coupling makes use of
hardly accessible reagents, etc.) Much more suitable are reactions involving
nucleophilic nitrogen atom (formation of amides, sulfamides, hydrazides, or
hydrazones), the best of which is the formation of compounds with an aminoalkyl
group. Substituted aminoalkyl group ($NHRCH_2-$ or NR_2CH_2-) is a well-known
pharmacophore group in a great variety of medicines and natural compounds. There
are several strategies for introducing this group into a desired substrate
(Scheme 4.1): Mannich reaction (1), *N*-alkylation (2), and reductive amination (3).

The first two reactions have natural limitations, since not every substrate can be
introduced into the Mannich reaction and not any combination of amine and
alkylating agent results in unambiguous monoalkylation. The third strategy (conversion of carbonyl group into aminoalkyl) is the only one which is practically
universal in nature, as reflected in detail in a series of reviews [1]. The popularity of
this reaction is largely associated with the wide distribution and accessibility of
aldehydes as there are numerous ways to introduce aldehyde group into a great
variety of substrates. The key feature of reaction (3) is that it involves intermediate
formation of imines (Schiff bases). This stage is reversible.

The reaction can be brought to completion to obtain the final amine in high yield
(Scheme 4.2), provided imine has been selectively reduced into amine (in the
presence of aldehyde). In the reaction of aldehydes with secondary amines,

© The Author(s) 2017
E.V. Babaev, *Incorporation of Heterocycles into Combinatorial Chemistry*,
SpringerBriefs in Molecular Science, DOI 10.1007/978-3-319-50015-7_4

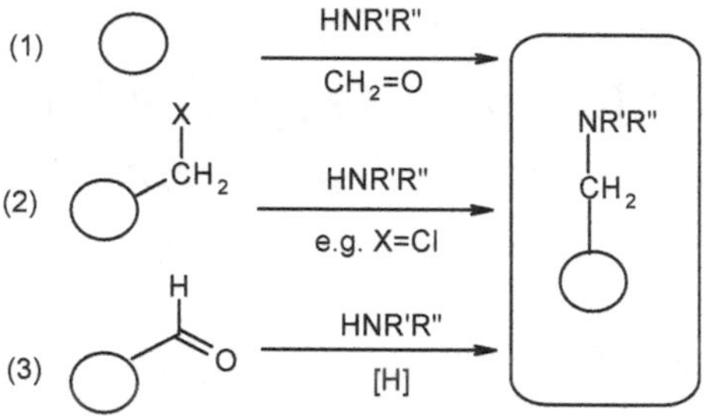

Scheme 4.1 Mannich reaction *1*, *N*-alkylation *2*, and reductive amination *3* leading to aminoalkyl substrates

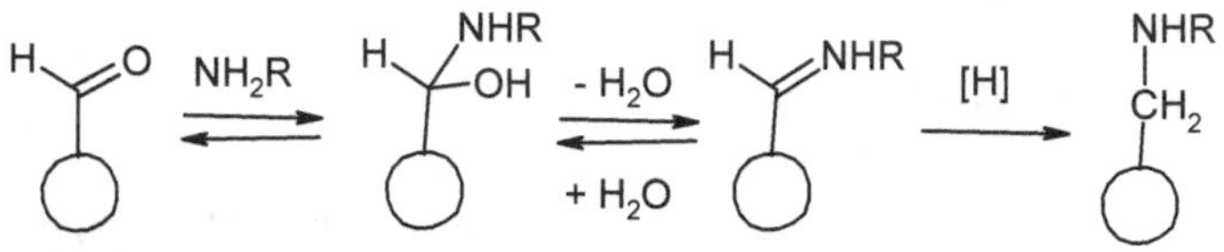

Scheme 4.2 The mechanism of reductive amination

reduction is experienced by the equilibrially formed iminium salt. The problem is to properly choose a selective reducer which reacts fast with imine and not slowly reacting with aldehyde [2]. The best reagents [3] were found to be Na[BH(OAc)$_3$] and Na[BH$_3$CN]. The standard admixtures in this reaction are traces of alcohol (side reduction of aldehyde), as well as unreacted starting materials. The reaction occurs under heterogeneous conditions in DCM or DCE, and efficient agitation of the heterogeneous mixture is the principal condition for complete reaction.

Products can be purified by using the simplest scavenger, specifically a Dowex ion-exchange resin. This resin is a styrene–divinylbenzene copolymer containing sulfonic acid groups which bond amines into salts. As a result, aldehyde is filtered through the resin and removed with the filtrate. The amine bound by the cationite is expelled from the resin by another amine, say, diethylamine (in methanol). It is quite clear that the final product (amine 2) of the "aldehyde + amine 1 = amine 2" reaction cannot be purified from the starting reagent (amine 1) in this way. The solution eluted from the resin necessitates further purification.

Having called the reductive amination reaction the most "combinatorial," we would like to focus on certain details of its implementation in the parallel mode. Such details are rarely described in scientific publications. At the same time, in Russia, this reaction has received a great deal of attention: The reaction was one of the most popular objects at Moscow research laboratories of the ChemBridge Company. As a result, ten (if not hundreds) thousands of compounds for testing comprise just this structural motif in the global market. Owing to the facility for the implementation, this reaction is convenient to use for teaching of the basic principles of liquid-phase parallel synthesis. It does not require sophisticated equipment and can be used even at low-budget training laboratories.

For this reason, the present paper is built as a training task (with protocols and instructions for students), as it was suggested to students of the Moscow State University over many years. The task is given in three versions. The first version is the simplest and requires virtually no equipment. The second version is based on a real practice of the implementation of this work at ChemBridge laboratories by use of interesting separation techniques. Finally, the third version represents the adaptation of this task to SynCore apparatus, performed at the MSU by the request of the producer company (Buchi). This synthesis was highly appreciated by the Swiss company, and our procedures were included in the "Best at Buchi" booklet [4]. The reactions performed at the MSU by the simplest, first scheme and subsequent in vivo tests resulted in the discovery of a new family of anxiolytics [5].

4.2 Synthesis Without Special Equipment

The training reductive amination reaction (like the Ugi reaction, see below) is a facile and an illustrative example highlighting specific features of parallel organic synthesis. The suggested training task was perfected by a group of students during the special laboratory course in combinatorial chemistry at the MSU in mid-2000s. In its chemical essence, it is a one-stage reductive amination reaction between series of heterocyclic aldehydes and aliphatic amines, with $Na[BH(OAc)_3]$ as the reducer (Scheme 4.3). The task involves the stage of parallel purification of products by means of the simplest scavenger, viz. an ion-exchange resin.

Each student performed six syntheses of four target compounds by a reductive amination (1 aldehyde + 4 amines), and, therewith, in three cases, the aldehydes/amine molar ratio was varied. Over the course of six lessons (2–3 h each), each student performed six parallel reactions, preliminary purification of the reaction product on a Dowex cation exchanger, and chromatographic purification of products, after which their NMR were measured. At the final lesson, students had to draw conclusion of the optimal reaction conditions and analyze the yields and purity of the products.

Reagents and Solvents: (1) Dichloroethane (DCE), dichloromethane (DCM), isopropanol, potassium carbonate, anhydrous sodium sulfate, silica gel for chromatography, and Silufol plates; (2) $Na[BH(OAc)_3]$ as the reducer and anhydrous oxalic acid for precipitating oxalates; (3) Dowex ion-exchange resin for purification of products from aldehydes admixtures; (4) a series of 5–10 secondary aliphatic amines, for example, cyclic (pyrrolidine, piperidine, and morpholine). We

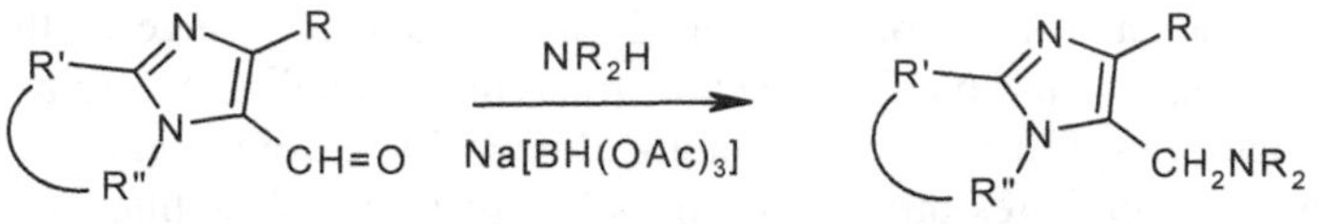

Scheme 4.3 Examples of heterocyclic aldehydes taken for reductive amination

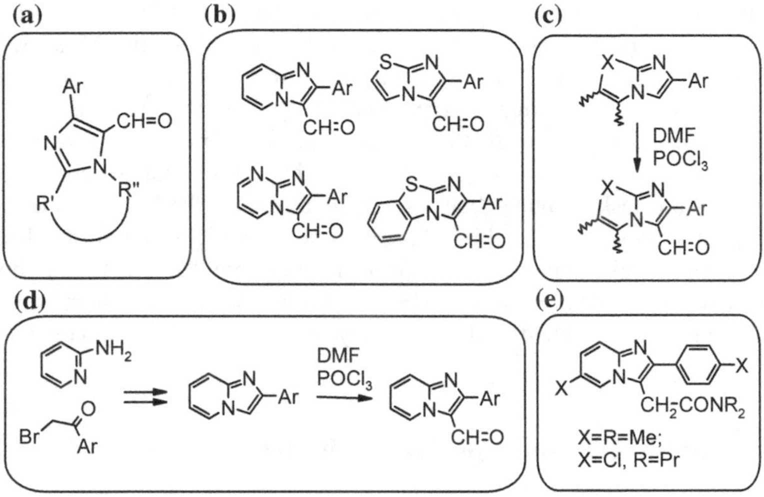

Scheme 4.4 General (**a**) and concrete (**b**) structures used, their preparation (**c**, **d**) and related drugs (**e**), see text

additionally used a series of amines on the basis of 4-substituted piperazines (4-alkyl and 4-acyl derivatives) and piperidine-4-carboxamides (substituents in such amines can be readily varied, provided 1-Boc-piperazine or N-Boc-isonipecotic acids are available in sufficient amounts); (5) a series of 5–10 aromatic aldehydes. We used a specific set of heterocyclic aldehydes of the imidazole series of general formula shown in Scheme 4.4a, in which R′ and R″ form a different heteroring: pyridine, pyrimidine, thiazole, or benzothiazole (Scheme 4.4b). Such aldehydes can be easily prepared by the Vilsmeier reaction by the formylation of the corresponding condensed imidazoles (Scheme 4.4c).

The starting bridged heterocycles are quite easily prepared by the classical Chichibabin scheme from aminoheterocycles and phenacyl bromides (the example for one subclass is given in Scheme 4.4d). The yields at all the stages are fairly high, and the procedures for the synthesis of aldehydes [6] and their precursors [7] can be found in the literature. As a result, the short synthetic sequence leading to aldehydes allows flexible variation of the heterocyclic and aryl residues in the aldehyde. The compounds formed by the reductive amination reaction are certainly structurally related (drug likeness) to known drugs—zolpidem and alpidem (Scheme 4.4e).

Equipment: (1) a set of automatic one-channel pipettes of various volumes with a set of tips for them; (2) a set of glass vials with screw caps (three times the number of reactions) and any shaker for agitation; (3) a set of plastic syringes with a porous diaphragm (Fig. 4.1) for passing reaction mixtures through the cation-exchange resin. Syringes can also be used as mini-columns for chromatography, but it is better to pass the mixtures through a short silica bed on a glass filter. One should think out thoroughly how to fix, in a convenient and space-saving way, six syringes

Fig. 4.1 Space-saving fastening of syringe columns in a manifold

(filters) at one workplace; for example, a simple metal manifold can be used (Fig. 4.1); (4) standard glassware with a great number of glasses (60–100 ml).

Acquired Skills. At the lessons, students get familiar with details of the reductive amination reactions and optimize its conditions; they master the techniques in parallel synthesis, purification, and in working with scavengers. To get familiar with the technological aspects of parallel synthesis, it is useful to repeat the same task in the plate mode.

Lesson 1. *Reagent Loading*. The students get a variant of the task to obtain four target compounds by the reductive amination reaction. The requirement is to obtain three target compounds in similar conditions (1 aldehyde plus 3 amines) and one target compound at varied reagent ratios (the same aldehydes plus the fourth amine in 1.2:1, 1:1, and 1:2 ratios). An example of possible reagent loading is shown in Table 4.1.

The teacher explains that in this case (variant F), one and the same aldehydes 5 of the aldehydes set (code CO5) is used in all experiments, and it is reacted with four amines (codes который NR09–NR12). Solutions of aldehydes and amines are prepared and enumerated in advance, and therefore, students fulfill the following simple operations:

(1) Stick labels on each vial, according to the variant obtained. For example, 1F, 2F, 3F, 4F, 5F, 6F (F is a task variant and the figure is an experiment number).

(2) A 0.343 M solution of aldehydes in DCE (2 ml) is added to each vial with a pipette or, for poorly soluble aldehydes, 2 ml of DCE is added to a weighed sample corresponding to 0.69 mmol of aldehyde.

(3) A DCE solution of required amine is transferred to each vial. The solutions are prepared in advance. In experiment nos. 1–4, add each amine solution (1.66 ml)

Table 4.1 Reagent codes and amounts and product codes in variant F

Exp. no.	Reagent codes		Reagent amounts		Product codes
	Aldehyde	Amine	Aldehyde, ml	Amine, m	
1F	CO5	NR09	2.0	1.66	1F-CO5-NR09
2F	CO5	NR10	2.0	1.66	2F-CO5-NR10
3F	CO5	NR11	2.0	1.66	3F-CO5-NR11
4F	CO5	NR12	2.0	1.66	4F-CO5-NR12
5F	CO5	NR12	2.0	2.0	5F-CO5-NR12
6F	CO5	NR12	2.0	4.0	6F-CO5-NR12

to each vial (aldehyde:amine 1.2:1). In experiment no. 5, add 2 ml of a solution of the fourth amine to aldehyde solution (aldehyde:amine 1:1). In experiment no. 6, add 4 ml of a solution of the fourth amine to aldehyde solution (aldehyde:amine 1:2). Close the vials with screw caps and place into a shaker for agitation.

Due to the fact that reaching equilibrium between the reagents with intermediate imine formation takes some time, it is therefore recommended to finish the lesson at this stage. However, the best results are obtained if the reducer is added in 12–24 h.

Lesson 2. *Imine Reduction*. (4) Add 2 ml of a suspension of a DCM solution of Na[BH(OAc)$_3$] to each vial with a pipette. Screw the vials with caps untightly (hydrogen is evolved) and place into the shaker. The reaction is usually complete in 24–48 h.

Lesson 3. *Isolation of the Target Product*. (5) Add 10 ml of 20% aqueous potassium carbonate to each vial. Screw the vials with caps untightly and agitate in the shaker for 30 min.

(6) Prepare the six empty vials with the same labels as on the reaction vials. Withdraw from each reaction vial with a syringe, that is the lower (organic) layer and transfer into the empty vial. Extract the aqueous layer once with 5 ml of DCM. Combine the organic layers.

(7) Perform TLC of the organic layer, using the starting aldehydes and amine (eluent CHCl$_3$/MeOH, 8:1 or C$_6$H$_6$/EtOAc, 2:1), Fig. 4.2 as references. It is recommended to apply on the first plate three first reaction mixtures plus three starting amines and aldehyde, and on the second plate, the other three reaction mixtures plus the fourth starting amine, and aldehyde.

(8a) If no starting aldehyde is observed in the chromatograms, add to the solutions anhydrous Na$_2$SO$_4$ as a drying agent. Screw the vials with caps and leave overnight.

Lesson 4. *Product Purification*. (8b) If the starting aldehyde is still observed, further purification should be performed. The first method: Shake the organic layer with aqueous sodium bisulfate and repeat TLC. The second method: Apply the organic layer on a column (Fig. 4.3) packed with Dowex saturated with pyridine. The resin well absorbs the product but poorly adsorbs the starting aldehyde. Wash the column with methanol until the eluate no longer contains traces of aldehyde. Elute the final product with methanolic diethylamine. Evaporate the methanolic solution under a hood. Additional purification requires chromatography (see item 9).

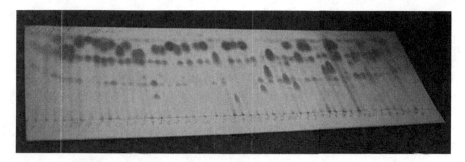

Fig. 4.2 Typical chromatogram after reductive amination (in UV light)

Fig. 4.3 Loading a suspension of an ion-exchange resin into columns with a dispenser and washing of the columns

(9) When the starting aldehyde is no longer present, but the starting amine is still present, chromatographic purification on silica (eluent CHCl$_3$/MeOH, 8:1) is required. As a rule, the starting aliphatic amine has low R_f, and therefore, it is readily separated by passing the mixture through a short silica bed on a glass filter.

(10) The final reductive amination products (after chromatographic purification) can be additionally purified by their precipitation as oxalates. To this end, add a solution of oxalic acid in isopropanol to product solutions and leave until precipitates form.

Lessons 5 and 6. *Filtration of Precipitates and Determination of Product Yields.*

(11) Filter off the oxalate precipitates, wash with ether, and dry.

(12) Stick labels with codes on the vials, for example, 1F-CO5-NR12. Here, 1F is the experiment number, CO5—aldehyde number, and NR12—amine number (Table 4.1). Weigh empty vials, determine the weight of products, and calculate the yields. Prepare one of the samples for ^{1}H NMR measurements. At the final lesson, the results of study of the effect of reagent ratio on product purity and yield are discussed. As a rule, the conversion rises if one of the reagents is present in excess, but you should "pay" for that by the necessity to remove excess aldehyde (scavenger) or amine (chromatography).

4.3 Technology of Microsynthesis in Plates

Here, special technologies and equipment are required to prepare large libraries of compounds by liquid-phase parallel synthesis. Such equipment is usually available in companies specializing in combinatorial synthesis.[1] In the ChemBridge laboratory, each student accomplished in a rack 48 reductive amination reactions (1 aldehyde with 48 amines). A total of 332 products were obtained. The following procedure is a protocol for fulfilling this task.

Practical implementation of the work

There are a number of simple techniques to facilitate liquid-phase parallel synthesis. First, metal block racks (6 × 8) for 48 tubes are used (Fig. 4.4). Reagents and solvents are added with multichannel pipettes (6 or 8 tips). The whole block is covered with a single Teflon seal, and the contact between the seal and tubes is improved by means of an additional metal screw top cover. Stirring is performed in a drum accommodating several blocks.

Stage 1. A plate for 48 glass tubes (6 ml) is charged with a suspension of finely ground Na[BH(OAc)$_3$] in DCM (800 μl, 3.6 mg, 0.17 mmol per tube; 3.6 g/40 ml CH$_2$Cl$_2$ per plate) by means of a Step-Pett dispenser. Solutions of corresponding amines (by 0.2 mmol) and aldehydes (by 0.26 mmol) in DCM (by 400 μl) are added to the reducer. If a different loading sequence is used, make sure there is no contact between aldehydes and reducer during loading. If a compound is poorly soluble, the solvent volume can be increased. Reagents are loaded with a 1000-μl pipette with Matrix long tips (1250 μl). The tips are positioned in an order corresponding to the loading order in a special stand and further used at all stages of isolation of the given load, after which the tips are washed in lines by means of an 8-channel pipette. The tubes are covered with a rectangular Teflon seal, fastened

---

[1]After fulfilling the previous task at the MSU, for students who desired to master the modern technology of parallel liquid-phase microsynthesis in plates, three lessons in the laboratory of the ChemBridge Company in Moscow were organized. As substrates, the same heterocyclic aldehydes as in the MSU were used, while the range of amines was essentially extended. Ten students took part in these lessons.

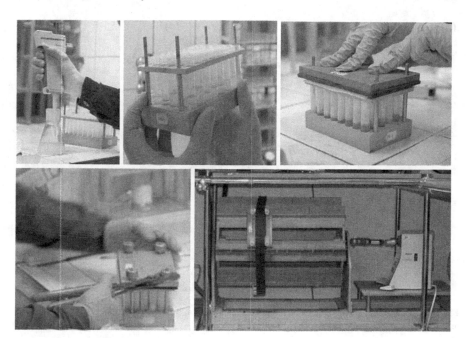

Fig. 4.4 Loading of reagents in the 6 × 8 format, hermetization of the plate, and parallel stirring of several plates

with an additional metal screw cover, and thoroughly mixed in a grill-type drum for 24 h. The solutions are reduced to 100 μl in air or in a drying oven at 40 °C.

The next stage involves separation of the final amine from unreacted aldehyde in each reaction mixture. To this end, like in the previous task, the mixture is passed through a Dowex cation-exchange resin. When working with 48 mixtures on columns, it is hard to follow the sorption process for the whole array of samples.

Stage 2. To remove aldehyde from the reaction mixtures, the solution residues obtained at the first stage are dissolved in methanol and transferred with an 8-channel pipette with 5-ml tips to columns loaded with 4 ml of a Dowex–pyridine suspension. (The suspension is prepared by magnetically stirring 90 g of Dowex and 100 ml of methanol see Fig. 4.5a.) The suspension is applied on the column with an 8-channel syringe dispenser or a 40-ml step dispenser (Fig. 4.5b). Further on, neutral admixtures (aldehydes) are eluted with methanol (5 ml per column) (Fig. 4.5c). The first 5 ml of the neutral layer is collected into 48 glass tubes (6 ml); therewith, the possible breakthrough of amine is followed by TLC. One more plate with resin columns is used, and the neutral layer is applied again if necessary. Methanol residues are pressed out from the columns with compressed air. The final products are eluted with a 30% solution of diethylamine in methanol into 48 glass tubes (6 ml) placed in two racks. Amine solutions should be poured into filtering columns as uniformly as possible so that all receivers are filled up to the top by the end of elution. If after addition of diethylamine the product starts to crystallize on

(a)   **(b)**   **(c)**

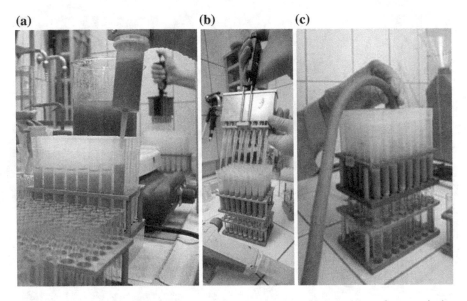

Fig. 4.5 Preliminary purification of 48 reaction mixtures on a Dowex cation-exchange resin (see text)

the resin, the methanol/diethylamine mixture (after the first 5-ml portion) should be replaced by a 1:1 chloroform/methanol mixture.

At the next stage, it is necessary to remove completely the solvent and diethylamine (bp 56 °C). To this end, a Savant vacuum centrifuge operated in the freeze-drying mode is used (see Chap. 3). The centrifugal force prevents solvent bumping, and even fairly high boiling solvents like DMSO can be evaporated.

Stage 3. The solvent is evaporated to dryness at 40 °C in a Savant centrifuge for 6 h, and the products are dissolved in 200–400 µl of chloroform and analyzed by TLC. The products which require no further purification are transferred into weighed plastic tubes (6 ml), labeled with barcodes for electronic data reading. Therewith, 10% of each solution is taken for NMR measurements and transferred to a Marsh plate (96 × 1.2 ml). Both plates are labeled (positions of unlabeled tubes correspond to the loading list), and the solvent is evaporated in air or in a drying oven at 40 °C, and removed completely in a vacuum drying oven. The loading list and chromatogram are archived. If, according to TLC data, the products contain traces of the starting amine, they are dissolved in 300 µl of chloroform and passed through chromatographic columns. Filtration is performed using Oros membrane-bottom plates (50 mg of silica, which corresponds to the mark in the lower part of the column).

Stage 4. Silica is charged with a well dispenser. To this end, silica is charged to a block with drilled-in holes and leveled for uniform distribution over holes (Fig. 4.6a). The holes are then superposed with wells (Fig. 4.6b), and the whole construction is turned upside down. As shown in (Fig. 4.6c), the sorbent is uniformly distributed over 48 (or 96) plate wells.

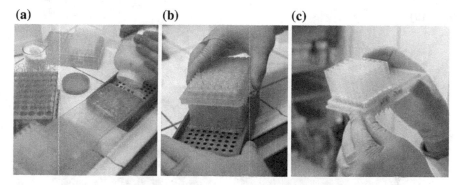

Fig. 4.6 Preparation of a plate for chromatographic separation of 48 compounds. For denotations, see text

Solutions are applied on columns with pipettes (Fig. 4.7a). Eluates are collected into 48 glass tubes (6 ml), silica is washed with chloroform (final volume of the eluate is ∼3 ml), and TLC analysis is performed. If necessary, the operation is repeated, after which the products are transferred into preweighed tubes. The products requiring preparative column chromatography (admixtures having higher R_f than the product) are collected separately for subsequent purification (Fig. 4.7b).

4.4 Liquid-Phase Parallel Synthesis in SynCore Apparatus

The next task was fulfilled using a SynCore apparatus for parallel synthesis. Our aim was to compare the work on high-cost equipment with the previous techniques: a "usual" laboratory technique and microsynthesis in plates. For the present task, we took a different series of heterocyclic aldehydes. Students reacted 5-arylfurfurals with secondary cyclic amines (Scheme 4.5). The acidophobic and oxidation-sensitive furan ring is considered as an "inconvenient" fragment in parallel synthesis, since furans are more reactive, and, therefore, give more by-products. The subclass of 5-arylfurfurals are more stable compounds obtained by arylation of furfural. The arylating agents are unstable diazonium salts which are readily available from various anilines. Therefore, it is easy to prepare a fairly wide series of aldehydes.

There are few examples of reductive amination of furfural and its aldehyde derivatives (including 5-aryl derivatives), but this subject has not yet been studied systematically. Interestingly, the resulting compounds contain a furfurylamine fragment which is a structural fragment of many drugs (the most known of them is furosemide). Therefore, we could consider these compounds as drug-like molecules (Scheme 4.6).

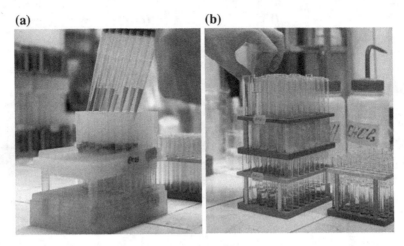

(a) **(b)**

Fig. 4.7 Parallel chromatography of 48 compounds (format 6 × 8). For denotations, see text

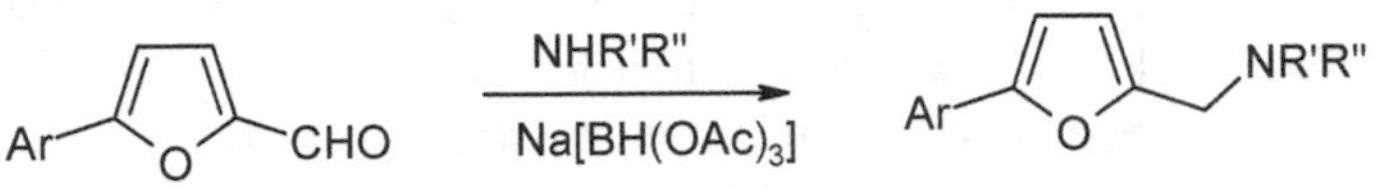

Scheme 4.5 Reductive amination of 5-arylfurfurals

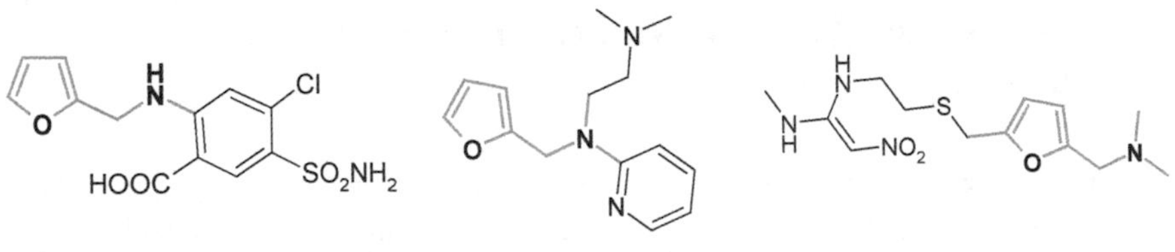

Furosemide
(diuretic)

Metafurilene
(antihystamine drug)

Ranitidine/Zantac
(stomach acid production inhibitor)

Scheme 4.6 Common drug prototypes with the furfurylamine fragment

Reagents and Solvents: (1) DCM, isopropanol, potassium carbonate, anhydrous sodium sulfate, silica gel for chromatography, Silufol plates; (2) Na[BH(OAc)$_3$] as the reducer and anhydrous oxalic acid for precipitating oxalates; (3) a series of secondary cyclic aliphatic amines (from 3–6 to 20) and a series of 5-arylfurfurals (from 4 to 10). The set is selected so that all 24 vessels of the apparatus are loaded. Combinations of 9 aldehydes and 18 amines (4 reactions for each student, a group of 20 students) are selected.

Equipment. The SynCore apparatus (modules for shaking, evaporation, and filtration) with 24 round-bottomed tubes (Fig. 4.8a, b). An ultrasonic bath (advisable) (Fig. 4.8d). Metal racks with various height 3 × 4 ports (Fig. 4.8c, d). The low rack serves as a stand can be placed as a whole into the ultrasonic bath

Fig. 4.8 **a** Disassembled and **b** assembled SynCore apparatus, and **c**, **e** additional equipment for performing the task

(Fig. 4.8c, d). The high rack is used for filtering precipitates or passing mixtures through filters with silica beds (Fig. 4.8e). Glasses (60–100 ml) and pipettes.

Acquired Skills. In this task, students get familiar with the SynCore apparatus which simultaneously fulfills the functions of stirring, shaking during extraction, filtration of 24 solutions from the drying agent, and analog of a rotor for evaporation from 24 vessels.

Practical implementation of the work

Each student gets a variant of the task to obtain four target compounds by reductive amination of one aldehyde with four different amines. Reagents and products are coded as above (Table 4.1). Reagents are loaded in a 1:1 ratio. The reducer is taken in a threefold excess.

Lesson 1. *Reagent Loading*. (1) Each tube is provided with a label specifying the obtained variant, for example, F-1, F-2, F-3, F-4 (F is task variant and the figure is an experiment number).

(2) Transfer 5 ml of a DCM solution of aldehyde (15 mmol) and then 2 ml of a DCM solution of corresponding amine (15 mmol) to tubes with a pipette. Place the tubes into SynCore for agitation.

Lesson 2. *Introduction of Reducer*. (3) Add to each tube with a pipette 3 ml of a DCM suspension of the Na[BH(OAc)$_3$]. Subject the tubes to ultrasound for 2–3 min for homogenization, then place them into SynCore, and activate the shaking mode. The best results are reached within 2–3 days (at 20 °C), but in rare cases, with poorly soluble compounds, the reaction may take up to 5 days. By the end of reaction, the reaction mixtures acquire a gelatinous consistency.

Lesson 3. *Treatment of Reaction Mixtures, Drying of Extracts*. (4) Add 20 ml of 20% aqueous potassium carbonate to each tube. Place the tube into SynCore and shake for 1 h. Over this period, hydrogen evolution ceases completely. Prepare four empty with the same labels as on the reaction tubes.

(5) Withdraw with a pipette or syringe the lower (organic) layer and transfer it to an empty tube. The remaining aqueous layer extracts two times with 10 ml of DCM. Combine the organic layers.

(6) Add anhydrous Na$_2$SO$_4$ to the organic layer (the drying agent should be added so that it does not fall on tube walls), shake the mixture for 30 min, and leave overnight.

Lesson 4. *Parallel Filtration from Drying Agent and Evaporation*. (7) Place the tubes into the reaction module of SynCore to filter off the drier. Place four pure tubes with labels to the receiver module in the strictly same order as in the reaction module. Perform parallel filtration and additional washing of the drier with a new solvent portion; to this end, attach the apparatus to a vacuum system (Fig. 4.9). The aim of this stage is to demonstrate to students an efficient system of parallel purification and washing of 24 reaction mixtures and filters from an inorganic dryer and from organic components. Full washing of the system (tubes and filters) is reached within 40 min by successive passing of two portions of water (2 × 30 ml) and acetone (2 × 30 ml). Then, the system is dried with a stream of nitrogen.

(8) Place the receiver module with collected filtrates into the SynCore rack, seal with a cover, attach to a vacuum pump (Buchi), and evaporate the solvent with shaking at 30–35 °C and a vacuum of about 300 mm to prevent bumping.

Lesson 5. *Chromatography and Purification*. (9) Perform TCL of the obtained residue using as references the starting aldehyde and amine (eluent CHCl$_3$/MeOH, 8:1, or C$_6$H$_6$/EtOAc, 2:1). If the starting aldehyde and/or amine is still present, perform chromatographic purification on a glass filter with silica bed (diameter 25 mm, bed height 30 mm), eluents chloroform and then CHCl$_3$/MeOH, 20:1.

(10) The reductive amination product can be further purified by its precipitation as oxalate. To this end, add to a CCl$_4$ solution of the product containing no starting amine admixture (or to a CHCl$_3$ solution after chromatographic purification) an isopropanol solution of oxalic acid (15 mmol), and leave the solution until precipitate formation. Filter off the precipitate through a usual glass filter and dry in air.

Lesson 6. *Determination of Product Yields*. (11) Prepare empty tubes and stick labels with product codes on them (the list of codes is available from the teacher).

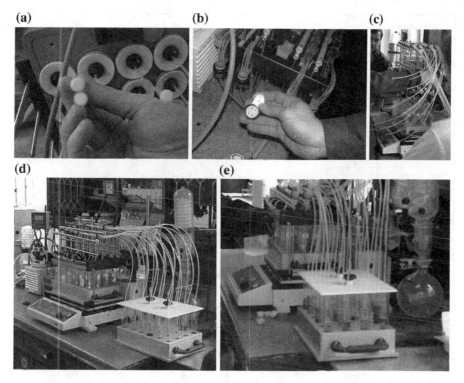

Fig. 4.9 Parallel filtering off of the drying agent in 24 samples: **a** cover with filters; **b** point of supply of the compressed gas to the distribution system pipes; **c** installation of the lid with the filters and tubes; **d** view of the instrument before starting the displacement solutions; **e** collection of the filtrates

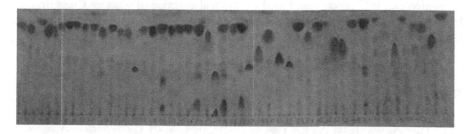

Fig. 4.10 Typical chromatogram after reductive amination of 5-arylfurfurals

Weigh the empty tubes. Determine the weight of products and calculate their yield. Provide one of the samples for ^{1}H NMR measurements.

The results of this work show that reductive amination in the 5-arylfurfural series is a reliable method of synthesis of substituted furfurylamines. In this reaction, unlike the above two tasks, further purification is required in very rare cases, Fig. 4.10. Therewith, the yields are generally quite high (Fig. 4.11).

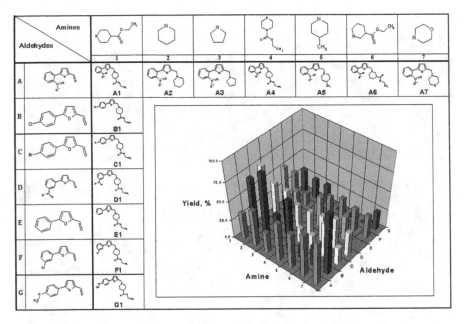

Fig. 4.11 The yields of reductive amination of 5-arylfurfurals. From the report in booklet [4]

4.5 Conclusions

The reductive amination reaction is rather convenient to use for teaching of the basic principles of liquid-phase parallel synthesis. Since it does not require sophisticated equipment, it can be used at low-budget training laboratories. Therefore, the task was given in three versions: the first one, that requires virtually no equipment, the second one, based on the real practice of work at ChemBridge laboratories, and the third one, which represents the adaptation of the same task to SynCore machine.

References

1. (a) Emerson WS (1948) Org Reactions 4:174–225. (b) Lane CF (1975) Sodium cyanoboro-hydride – a highly selective reducing agent for organic functional groups. Synthesis 135–146. doi:10.1055/s-1975-23685. (c) Hutchins RO, Hutchins MK In: Comprehensive Organic Synthesis (1991) Trost BM, Fleming I (eds) Pergamon, Oxford vol 8, p 25. (d) Baxter EW, Reitz AB (2002) Org Reactions 59:1–714
2. Henkel T, Brunne RM, Mueller H, Reichel F (1999) Statistical investigation into the structural complementarity of natural products and synthetic compounds. Angew Chem Int Ed 38:643–647. doi:10.1002/(SICI)1521-3773(19990301)38:5%3C643:AID-ANIE643%3E3.0.CO;2-G
3. Abdel-Majid AF, Carson KG, Harris BD, Maryanoff CA, Shah RD (1996) Reductive amination of aldehydes and ketones with sodium triacetoxyborohydride. studies on direct and indirect reductive amination procedures. J Org Chem 61:3849–3862. doi:10.1021/jo960057x

4. Babaev E, Belykh E, Dlinnykh I, Tkach N, Bender W, Shoenenberger G (2004) Parallel synthesis – reductive amination of aldehyde group. Best@Buchi Synth: 34
5. Geronikaki A, Babaev E, Dearden J, Dehaen W, Filimonov D, Galaeva I, Krajneva V, Lagunin A, Macaev F, Molodavkin G, Poroikov V, Pogrebnoi S, Saloutin V, Stepanchikova A, Stingaci E, Tkach N, Vlad L, Voronina T (2004) Design, synthesis, computational and biological evaluation of new anxiolytics. Bioorg Med Chem 12:6559–6568. doi:10.1016/j.bmc.2004.09.016
6. Saldabol NO, Popelis YY, Slavinskaya VA (2001) Formylation of furyl-substituted imidazo[1,2-a]pyridine, imidazo[1,2-a]pyrimidine and imidazo[2,1-b]thiazole. Chem Heteroc Comp 37 (8):1021–1024. doi:10.1023/A:1012799920436
7. Tschitschibabin AE (1926) Zur Tautomerie des α-Amino-pyridines, IV. Mitteilung Ber 59:2048–2055

Chapter 5
A Parallel Ugi Reaction

In the beginning of 2000s, at the Ural State Technical University, Yekaterinburg, there was developed a "Combinatorial Chemistry" laboratory training course for students specializing in biotechnology (Dr. M. Mironov). The aim of this laboratory course was to develop practical skills in the field of parallel synthesis and consolidate the lecture material on combinatorial chemistry. Taking into account the basic student' training level, we had to find a simple and illustrative example of the parallel organic synthesis technology. The Ugi reaction is a good example, since it allows compound libraries to be obtained without complex equipment and hardly accessible reagents.[1] The reagents were chosen to obtain precipitates which could be separated in parallel by centrifuging. The practical work at the USTU was envisioned for four lessons: an introductory workshop and three practical lessons, and an extended program comprising five practical lessons requiring a higher level basic training was suggested for year four students. In the first part of this paper, we present a series of procedures and instructions which can help one to include the Ugi reaction in any laboratory training in combinatorial chemistry.

Almost simultaneously, special laboratory training in combinatorial chemistry for fifth year students was initiated at the Lomonosov Moscow State University. The Moscow's group first experimented with the training tasks on solid-phase synthesis, and the task in parallel liquid-phase synthesis involved reductive amination. After the heads of these training courses had met face-to-face and exchanged experience, the Moscow group attempted to adapt the procedure for use for the SynCore reactor. In the second part of this paper, we present the results of this experiment. In our opinion, the experience in such collaboration may be of interest for other universities.

---

[1]It should be mentioned that the Ugi reaction is also used at students' laboratories at the Munich Technical University.

© The Author(s) 2017
E.V. Babaev, *Incorporation of Heterocycles into Combinatorial Chemistry*,
SpringerBriefs in Molecular Science, DOI 10.1007/978-3-319-50015-7_5

5.1 Basic Information on the Ugi Reaction

We consider it useful, to start with a workshop on the methodology of parallel synthesis and application of multicomponent reactions (MCRs). The term MCR relates to reactions that occur on direct mixing of three and more reagents, and the final structure includes fragments of all starting reagents. There are different classifications of MCR, depending on their mechanistic features [1]. The main advantage of such reactions is that they allow one to obtain a great number of derivatives in one stage from simple and accessible starting materials. It should be noted that MCRs are widely used for solving diverse practical tasks in the search for new biologically active compounds, catalysts, new materials, etc. One of the most popular of reactions of this type is the four-component Ugi condensation discovered in 1960. The reaction involves isocyanides, carbonyl compounds, amines, and organic/inorganic acids (Scheme 5.1).

The Ugi condensation provides a great number of derivatives. It features a considerable flexibility: Presently, several tens of its versions are known which open up the way to various scaffold structures. Finally, it has mild reaction conditions (room temperature), which is ideally suited for automated synthesis. All these advantages attracted attention of researchers working in combinatorial chemistry, as well as pharmaceutical companies searching for new biologically active compounds. At present, abundant published material on the reaction itself and practically valuable compounds obtained by this reaction is available. In preparation to the workshop, we recommend the reviews [2].

5.2 Experience in Accomplishing the Ugi Reaction Without Special Equipment

As the training task for students, we use one of the simplest versions of the Ugi reactions, specifically preparation of amino acids by mixing an aromatic isocyanide, aliphatic ketones, benzylamines, and phthalyl glycine (Scheme 5.2). This choice is not occasional; it allows one to perform a practical work with minimal preparation, provides excellently reproducible results, and requires no expensive or dangerous reagents. Use of phthalyl glycine (that gives poorly soluble derivatives) makes possible to readily crystallize the products which can be isolated in parallel in the

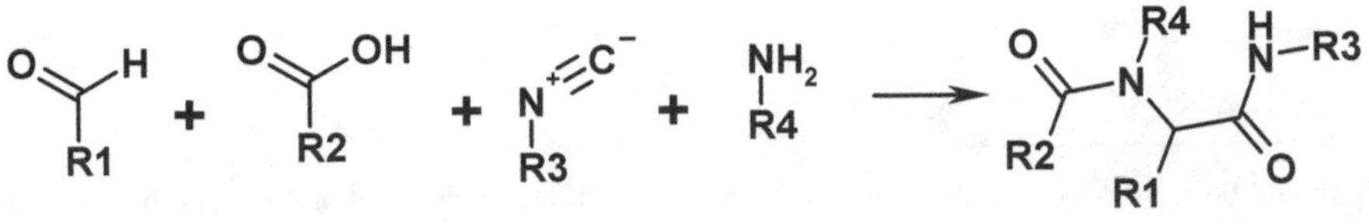

Scheme 5.1 Scheme of the Ugi reaction

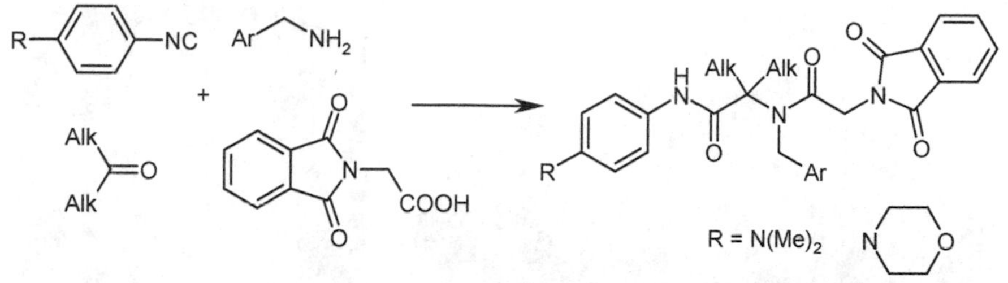

Scheme 5.2 Example of the Ugi reaction elaborated for students

framework of one laboratory work. In their turn, aromatic isocyanides with polar groups are low volatile and have almost no characteristic odor. Benzylamines and aliphatic ketones whose function is to provide various side-chain substituents are accessible and fairly cheap reagents. It should be noted that the described laboratory course was developed for engineering students and envisioned no subsequent testing of the resulting compounds. Therefore, the scaffold in Scheme 5.2 does not have biologically active analogs. For students of other specializations (e.g., medicinal chemistry), one can use a different target compounds, according to [2]. The procedures described below all remain therewith unchanged.

Practical skills acquired during the work. Students use equipment for parallel synthesis: single- or multichannel micropipettes, reaction plates, and shaker. A short introductory lesson is required if the students have never dealt with such equipment. The acquired skills may prove useful at microscale divisions of pharmaceutical and chemical companies, as well as at any analytical laboratory.

Materials and instruments. For laboratory work, the following equipment should be prepared: (1) plastic 8 × 12 plates with separate tubes (each 1 ml in volume), such as Micronic or Falcon plates (Fig. 5.1a). Well plates used in analytical chemistry are less convenient. (2) Fixed-volume 50- or 100-μl micropipettes (both single- and multichannel, Fig. 5.1b), as well as tip sets for them. (3) Shaker (almost any model can be used). (4) Laboratory centrifuge with a swing-bucket rotor. Centrifuges with an angle rotor are less convenient. (5) Set for thin-layer chromatography or analytical chromatograph. (6) Glassware for starting solutions, balances, and other laboratory equipment.

Reagents. For charging one plate, one needs about 500 mg of each reagent.

(1) Phthalyl glycine is prepared by fusing equimolar amounts of phthalic anhydride and glycine in a porcelain dish to obtain a homogeneous mass which is then crystallized from aqueous alcohol, mp 198 °C [3].
(2) Isocyanide, for example, 4-(dimethylamino)-phenyl isocyanide, is prepared by the procedure in [4]. A suspension of 3.2 mmol of 4-(dimethylamino)-nitrosobenzene in 40 ml of ethanol is added to a solution of 3.2 mmol of 3-phenylisoxazol-5-one (can be readily synthesized from benzoylacetic ester and hydroxylamine) in 20 ml of ethanol. The mixture is heated on a water bath for 15 min, cooled, and the precipitate is filtered off. The product is heated in

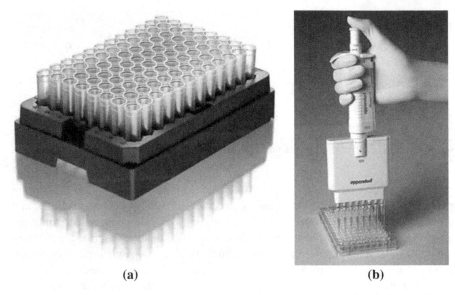

(a)   (b)

Fig. 5.1 Equipment for the implementation of the Ugi reaction. **a** Plastic plates and **b** multichannel pipette

toluene at 90 °C for 30 h, toluene is removed by distillation, and the residue is sublimed in a vacuum at 75 °C. Total yield is 65–70%, mp 61 °C. This reagent can be replaced by other aromatic isocyanides: 4-morpholinophenyl isocyanide or 2,4-dimethoxyphenyl isocyanide. The advantage of these reagents is their high activity and lack of the characteristic isocyanide odor.

(3) Substituted benzylamines, ketones (Scheme 5.3), and methanol. (If methanol is prohibited for use, it can be replaced by a 4:1 acetonitrile–water mixture.)

5.2.1 Practical Implementation of the Work

Lesson 1. *Reagent Loading.* Each student or a group of students obtain a hard copy of the database corresponding to the library to be obtained. Before starting experimental work, students should write down equations of the reactions leading to the target structure. At the colloquium, if it precedes the work, detailed analysis of the mechanism of the Ugi reaction can be given. Then, students form in themselves, a reagent loading table in which isocyanide and acid are invariable and benzylamines and ketones are variable parameters. Therewith, benzylamines are labeled with digits and ketones, with letters (Scheme 5.3).

The next step is to prepare solutions of the starting compounds. The concentration of all solutions is 1 mmol ml^{-1} in methanol. A total of 12 solutions are prepared: 1 with isocyanide, 1 with phthalyl glycine, 4 with ketones, and 6 with

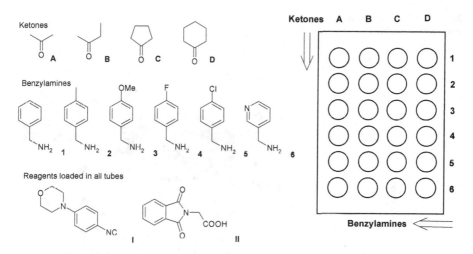

Scheme 5.3 Reagents used for implementation of the Ugi reaction

benzyl amines. The choice of ketones and benzylamines depends on the potential of each concrete laboratory. According to our observation, virtually any set allows amino acid derivatives to be prepared with good yields by the below-described procedures. As a recommendation, we provide the set shown in Scheme 5.3. Furthermore, students start to load all starting reagents, by 0.1 mmol each, into plates by means of 100 μl (50 μl, if the concentration of all solutions is 2 mmol ml^{-1}).

The following loading sequence is strictly followed: (1) benzylamines, (2) ketones, (3) isocyanide, and (4) phthalyl glycine. It should be noted that only part of the plate (4 × 6) is loaded, and therewith, loading is better started from the left upper angle (tube A1). All manipulations with solutions are performed under a hood. If students have never dealt with micropipettes, a preliminary practical lesson is recommended. It is useful to record the time of loading with multichannel pipettes to estimate the time gain compared with traditional methods. When loading is complete, plates are covered, transferred to a shaker, and agitate at room temperature for 20 min. The students should visually estimate the efficiency of mixing.

Lesson 2. *Product Isolation.* In the case of correct loading, all target compounds precipitate. The reaction time varies from 4 to 6 h, depending on a concrete reagent set. Therefore, the second practical lesson is recommended to be held on the next day. (Note that prolonged keeping results in decreased product yields.) Before the lesson, students should be instructed on how to operate the centrifuge and to discuss its operation principles. Then, students prepare a washing liquid (ethanol–water, 3:1). Further actions are determined by the type of the centrifuge, and the best centrifuge here is a centrifuge equipped with a special rotor for microplates. However, almost any model with the rotation speed 3000–4000 rpm (Fig. 5.2) is suitable. Wet precipitates in the plate are placed in an oven (<50 °C) and dried for 5–6 h.

Fig. 5.2 Centrifuge with a
rotor for microplates

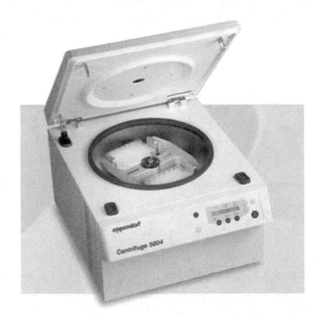

Table 5.1 Yields of target
products (%), obtained by a
group of students (loading by
Scheme 5.3)

		Benzylamines					
		1	2	3	4	5	6
Ketones	A	49	48	41	55	62	46
	B	54	67	60	65	73	51
	C	76	81	66	68	77	58
	D	78	75	62	69	71	53

Lesson 3. *Product Analysis.* If all the above-described operations are fulfilled
properly, the purity of products is higher than 90%. Since the reaction proceeds
sufficiently completely in all cases, the yields of products depend, first of all, on
their solubility in the washing liquid. The parallel liquid-phase technology presented in our example envisions the same is procedures for isolation of all target
compounds, which makes possible high isolation rates but does not ensure high
yields (Table 5.1).

The students weigh the precipitates and determine the yields of the target
compounds, which should fall into the range 40–80%. Then, the students analyze
the target products by TLC (an analytical chromatograph is also suitable). For TLC,
the samples are dissolved in chloroform and chromatographed in CHCl$_3$/ethanol
(95:5). The chromatograms are observed under a UV light, and R_f values are
recorded. Reaction completeness is estimated by the presence or absence of the
isocyanide spot. The samples synthesized by a previous group of students can be
used as references. A more detailed analysis of admixtures is performed by ^{1}H
NMR (Figs. 5.3 and 5.4).

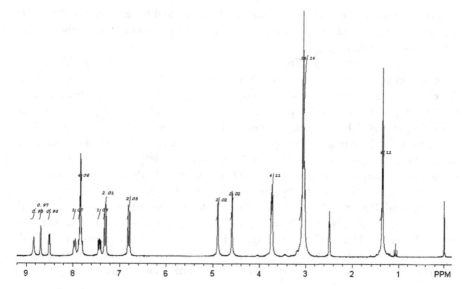

Fig. 5.3 ¹H NMR spectrum of compound **A6**

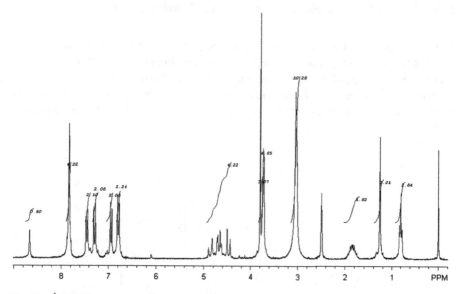

Fig. 5.4 ¹H NMR spectrum of compound **B3**

As shown in Table 5.1, lower yields are observed for compounds containing polar groups like pyridin-3-yl or 4-methoxyphenyl. It was found that the target compounds are partly lost with the washing liquid. The water/ethanol ratio in the washing liquid affects the yields and purity of the resulting products. For better results, this ratio can be varied to obtain the highest yield for each series. ^{1}H NMR spectra were measured for the compound obtained by students. No further purification was performed.

Compound **A6**. mp 275–276 °C, purity >95%. The spectrum (Fig. 5.3) shows signals of ethanol, implying incomplete drying. The spectrum of **A6** is typical for the entire series of compounds. The side-group signals are readily identifiable: 2 methylene groups at 4.5–5.0 ppm, 2 methyl groups at 1.0–1.5 ppm, and four pyridyl signals at 7.4–8.8 ppm. The phthalyl fragment signals are observed at 7.8–8.0 ppm as a four-proton multiplet. The phenylmorpholine fragment gives a characteristic 4-signal set (2 signals of the aromatic moiety at 6.6–7.4 ppm and 2 signals of the aliphatic moiety at 3.0–3.7 ppm), one of which overlaps with the water signal. The NH proton appears as a downfield signal at 8.8–9.2 ppm.

Compound **B3**. mp 280–281 °C, purity ∼90%. The admixtures (see the ^{1}H NMR spectrum in Fig. 5.4) here are the starting compounds: isocyanide and phthalyl glycine. The signals are assigned by analogy with the above compound. The difference consists in that the methylene proton signals are non-equivalent, which is the characteristic of derivatives of unsymmetrical ketones.

5.3   Variant of Implementing the Task with SynCore

The given task was first suggested to students of the Chemical Department of the MSU at the special laboratory course in combinatorial chemistry in 2006. The work was performed in SynCore apparatus for parallel synthesis.

Reagents, materials, and instruments are: phthalyl glycine, 4-(dimethylamino) phenyl isocyanide, a series of the simplest ketones (Scheme 5.4), and the SynCore apparatus.

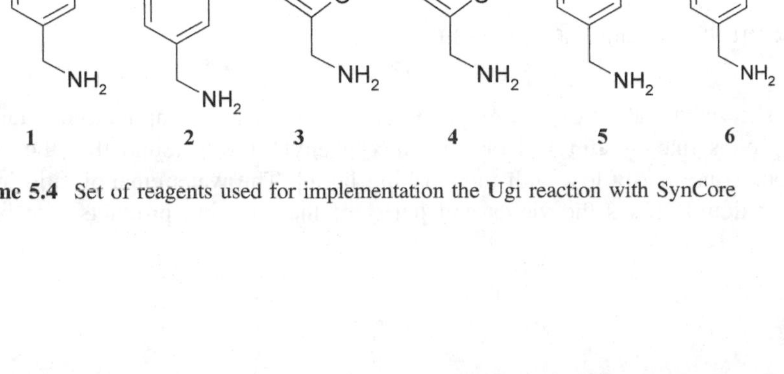

Scheme 5.4 Set of reagents used for implementation the Ugi reaction with SynCore

5.3.1 Practical Implementation of the Task

The students get a variant with the task to synthesize one compound by a multi-component Ugi reaction. All starting reagents are loaded in equimolar amounts (1.0 mmol).

Lesson 1. *Reagent Loading.* A special tube for SynCore is provided by a label corresponding to the variant obtained. For example, **B-3** (**B** is ketone and **3** is benzylamine). Each tube is loaded, in strict sequence, with 1 ml of an alcohol solution of amine, 1 ml of a ketone solution, 2 ml of an isonitrile solution (in a hood), and 1 ml of a phthalyl glycine solution. The tube is then placed into the reaction module of SynCore and agitated for 4 h.

Lesson 2. *Product Isolation.* The precipitates are filtered off in the filtration block of SynCore (Fig. 5.5a) and washed two times with alcohol/water (3:1) in the parallel mode (Fig. 5.5b), after which the tubes with precipitates were placed into a vacuum drying oven. Unlike the reductive amination task (where on filtering on SynCore, we collected mother liquors) in the given task, we collect and separate precipitates remaining in the reaction vessels. On filtering (Fig. 5.5a), a stream of nitrogen is fed into all the 24 tubes. As a result, the mother liquor is forced through 24 plastic tubings. These tubings are then replaced with hermetic seals, and the system is attached, by a common plastic tubing, to a flask with the washing liquid (Fig. 5.5b). To wash all the 24 precipitates simultaneously, a rubber tubing is attached to vacuum, and the washing liquid from the flask is admitted uniformly into all the 24 tubes. In necessary, the shaker is turned on, and the precipitates are suspended. The filtration procedure is repeated the required number of times.

Lesson 3. *Assessment of the Yield and Purity of Products.* A clean vial is provided with a label with the task variant and weighed. The precipitate is then transferred into this vial, the latter is weighed again, and the yield is estimated. The

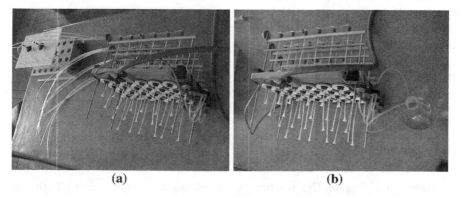

(a)   **(b)**

Fig. 5.5 Attachment for SynCore apparatus for parallel filtration of 24 suspensions: **a** for filtration (24 plastic tubings are attached to the *left* part of the attachment for forcing over the mother liquors) and **b** for washing (a bottle with a washing liquid is attached to the *right* part of the attachment)

Table 5.2 Yields of target products in the Ugi reaction (%)

Ketones	Amines					
	1	2	3	4	5	6
A	40	11	32	42	50	23
B	a	10	a	6	11	20
C	38	10	43	41	28	32
D	25	–	54	38	20	9

^aThe reaction was not performed

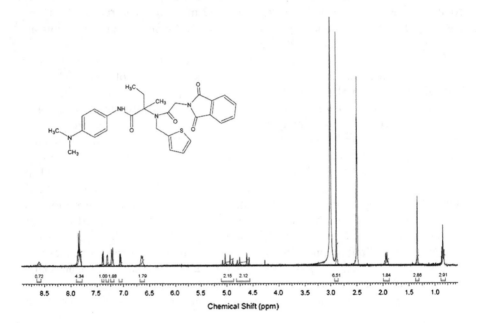

Fig. 5.6 ¹H NMR spectrum of one of the obtained products

purity of the resulting compound is determined by TLC, and samples for spectral analysis are taken. The percent yields of the Ugi reaction are listed in Table 5.2, and an example spectrum of one the products is shown in Fig. 5.6. Comparison with the data in Table 5.1 shows that the product yields are slightly lower and the product purities are roughly the same as in the original version.

5.4   Conclusions

In both approaches to Ugi reaction, we used poorly soluble phthalyl glycine derivatives and almost odorless isonitriles. These two features allow us to use Ugi reaction for demonstrations.

References

1. Multicomponent Reactions (2005) Zhu J, Bienayme H (eds) Wiley–VCH, Weinheim
2. (a) Ugi I, Dömling A (2000) Multicomponent Reactions with Isocyanides. Angew Chem Int Ed 39: 3168–3210. doi:10.1002/1521-3773(20000915)39:18<3168::AID-ANIE3168>3.0.CO;2-U. (b) Dömling A (2006) Recent Developments in Isocyanide Based Multicomponent Reactions in Applied Chemistry. Chem Rev 106: 17–89. doi:10.1021/cr0505728
3. Greenstein JP, Wintz M (1961) Chemistry of the amino acids. Wiley, New York
4. Wentrup C, Stutz U, Wollweber H-J (1978) Synthese von aryl- und heteroaryl isocyaniden aus nitrosoverbindungen. Angew Chem 90(9):731–732. doi:10.1002/ange.19780900925

Chapter 6
Combinatorial Heterocyclic Chemistry in Higher School

The progress of combinatorial chemistry in Russia is directly related to the problem of survival of domestic science (development of material basis, preservation of human resources), adaptation of basic science to practice, and search for optimum interaction between researchers and consumers of the final product. This is just the reason why the available experience in combinatorial R&D is not only of historical interest, but can also form the basis for further developments. The aim of the present chapter is to give a brief review of research, applied, and educational and managerial projects associated with combinatorial chemistry fulfilled over the past decade by our research group at the Chemical Department of the Moscow State University (MSU). We shall dwell on the experience in the implementation of fairly abstract synthetic ideas into practical projects of combinatorial chemistry.

6.1 Certain Applied Aspects of Heterocyclic Chemistry

Over a long time, our research efforts have been focused on bridged azolazines (Scheme 6.1). The parent structure in this family is indolizine, which can be considered as a superposition of pyrrole and pyridine, two key archetypes of heterocyclic chemistry as a whole. Indolizines are isomeric and isostructural to indoles (certain indolizines can even be rearranged into indoles), which induced out interest in this subclass. In developing synthetic approaches to functionalized indolizines, we discovered a number of new and quite unusual types of cyclizations and rearrangements.

The novelty was that the precursors of indolizines were cationoid or mesoionic azolopyridines, as well as pyridinium salts. Our long-standing research program [1]

© The Author(s) 2017
E.V. Babaev, *Incorporation of Heterocycles into Combinatorial Chemistry*,
SpringerBriefs in Molecular Science, DOI 10.1007/978-3-319-50015-7_6

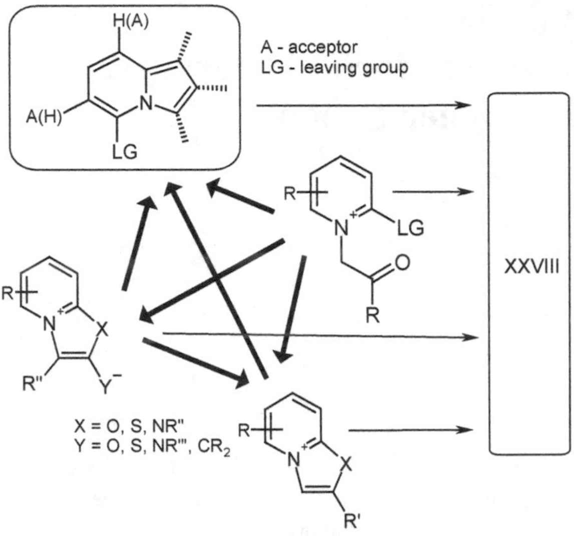

Scheme 6.1 General
interrelationship of
compounds studied

involved study of methods of the synthesis of compounds of these classes, as well as use of these compounds as precursors of heterocyclic systems **XXVIII**. Over the course of this research, we synthesized hundreds of new compounds belonging to previously unknown subclasses of substituted heterocycles.

Obviously, this research is indirectly related to the problems of combinatorial chemistry. Radically new compound classes could at best enrich abundant combinatorial collections of "historical libraries," where a new structural type would contribute to the diversity of structural motifs. (In this case, a new compound will probably be demanded in the framework of a narrow-focus project, for example, for testing on a concrete biological target.) As a result, the very novelty of the structural type is depreciated. It is good if structures of a new subclass are structurally similar to already known drugs. This can give guidance to what kind of activity is purposeful to test for initially. And what to do then with absolutely new compounds?

One of the few approaches that have been able to at least roughly answer this question is the program PASS [2]. A user enters the structural formula (standard mol/sdf file) of one or many compounds, and the program predicts their possible biological activity spectrum. The algorithm of this software can be briefly outlined as follows.[1]

In the PASS program, thousands of known drugs are already broken into small fragments; each fragment is assigned the same activity as in the whole molecule.

---

[1]Once the developers of the PASS program failed to attend some scientific conference and asked the author of this book to present their program instead of them. As a result, we had to explore in detail the algorithm of PASS.

If the same fragment is present in another drug (which has a different structure but exhibits the same type of biological activity), the weight factor of this fragment increases. If such fragments are contained in an unknown structure, the program estimates the probability of one or another kind of activity. In our research, we could obtain strong evidence for the predictions of the PASS program by directed biological testing of new families of compounds. Below, we consider some samples for concrete classes of compounds.

6.2   Synthesis of Antimicrobial Aminodienes

About ten years ago, we found that the pyridine ring of azolopyridinium salts opens under the action of amines to form substituted dienes **XXIX** (Scheme 6.2a) [3]. We considered this family to hold great promise for the synthesis of combinatorial libraries. The substituents (amine, aryl, and azole) and stereochemistry of the diene chain were readily varied, and the precursors of azolopyridinium salts were readily accessible compounds **XXX**.

At the same time, what kind of biological activity may be characteristic of such dienes was absolutely unclear, and we tried to predict it by means of the PASS program. The most probable was found to be the antimicrobial activity, and this prognosis was the same for the whole family of azoles **XXIX**. Our experiments with gram-positive (*Staphylococcus aureus*) and gram-negative (*Escherichia coli*) microorganisms showed that, too, all compounds exhibited activity against both types of bacteria (Table 6.1).

Later, we found that the same antimicrobial dienes of subclass **XXIX** can be obtained in high yields in a shorter way: by quite an unusual transformation of 2-halopyridinium salts **XXX** (Scheme 6.2b, Y = Cl) [4]. The mechanism of this elegant one-pot transformation of pyridines into oxazoles appears to be tandem in nature (oxazole closing + pyridine destruction), and it provides a facile and efficient synthetic route to a wide series of oxazoles **XXIXa**. It was established that salts **XXX** with Y = Cl can be replaced by their analogs with Y = MeS [5]. Such a replacement of function Y allowed us to realize a solid-phase modification of this reaction (Scheme 6.2c). As seen, the required pyridinium salt **XXXa** is "grown" through its sulfur atom on a solid support Z. As phase Z, we tried Merrifield resin and silica gel modified with a bromoalkyl residue. At the final stage, the sulfur-containing linker disappears entirely from the final structure (remaining on the solid phase under the action of amine), while a pure oxazole **XXIXa** passes to solution. Our approach clearly demonstrates the logics in the answer to the question, *what to do with new classes of compounds?* First, one should try to predict and search for their useful properties (Table 6.1). Second, one should optimize the synthesis of libraries by searching for new liquid-phase reactions (Scheme 6.2b) and still more efficient solid-phase approaches (Scheme 6.2c).

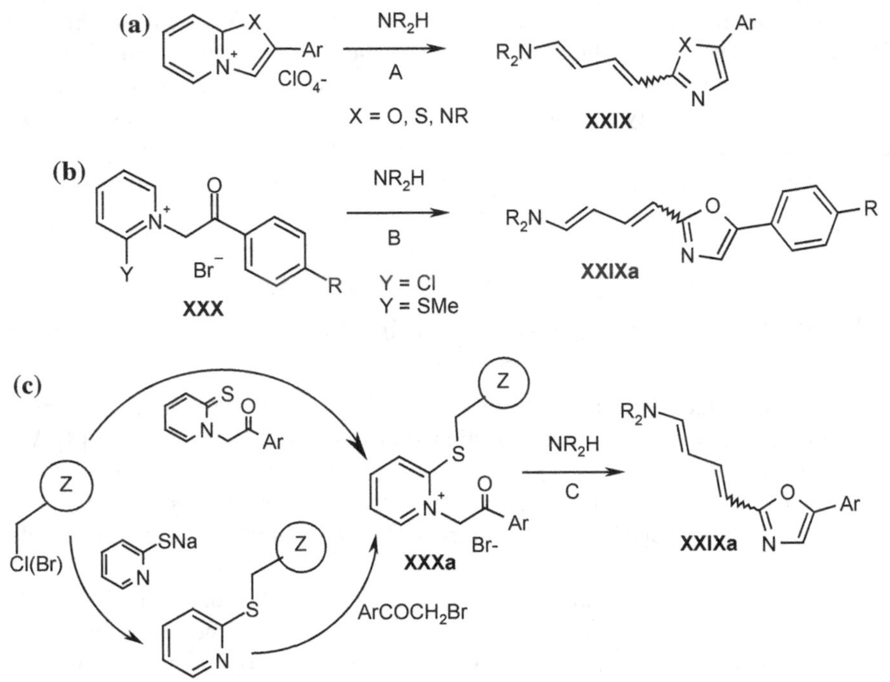

Scheme 6.2 Synthesis of azolyldienes **a** by ring opening of azolopyridinium salts and by ring transformation of the pyridinium salts: LPOS (**b**) and SPOS (**c**) protocols

Table 6.1 Antibacterial activity of compounds **XXIX**

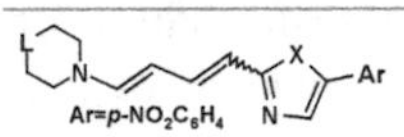	X	O	O	O	NMe	S
	L	CH$_2$	O	CH$_2$CH$_2$	CH$_2$	CH$_2$
S. aureus 6838		100	200	200	200	200
E. coli 25922		200	200	200	>200	>200

Minimum bacteriostatic concentration (μg ml^{-1})

6.3 Synthesis of 5-Aminoindolizines with Adrenergic Activity

Exploring reactions of amines with homologs of oxazolopyridinium salts, we unexpectedly obtained 5-aminoindolizines **XXXI** (Scheme 6.3) [6]. The possible intermediates of this quite a rare example of the conversion of the oxazole ring into pyrrole are shown in the bottom part of Scheme 6.3. We showed [1, 7] that this reaction is general in nature. One can attach alicycles to the pyridine fragment to introduce into it an electron-acceptor groups, and substitute the six-membered

Scheme 6.3 Discovered transformation of the oxazolopyridinium salts to aminoindolizines (*top*) and possible intermediates (*bottom*)

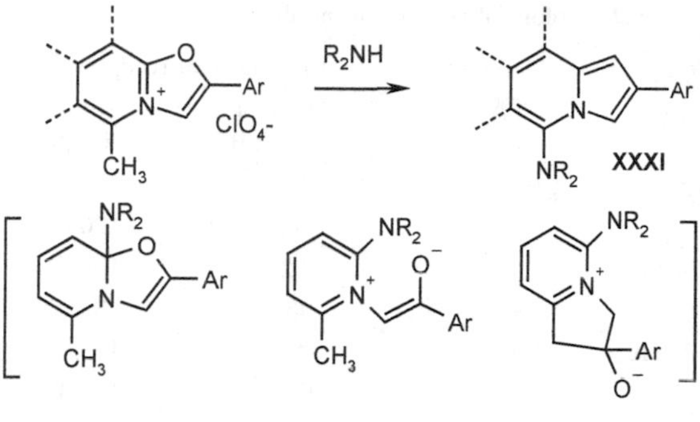

fragment by pyrimidine and secondary amines by alcoholates. The examples of structural classes of the resulting products are diverse, and as the starting compounds for preparing starting salts (and their analogs), we used readily accessible pyridin-2-ones and pyrimidinones.

Before our works, indolizines with donor substituents were unknown. As in the previous example with dienes, the type of biological activity of these compounds was unobvious (even though structurally compounds **XXXI** resembled their isostructural psychotropic indoles of the psilocin series). The PASS program predicted for indolizines XXXI a high probability of binding with β2-adrenoreceptors. Based on this prognosis, we selected perspective 5-aminoindolizine structures and tested them for binding to rat brain synaptosomal membrane receptors, by comparing their ability to expel tritium-labeled reference compounds (Table 6.2). The resulting data allow a conclusion that compound **XXXI** deserves further study with the aim to reveal its adrenergic activity. At present, antagonists of various subtypes of β-adrenoreceptors (practotol, atenolol, metoprolol, salbutamol, and soterenol) are widely used in clinical practice for treating a number of cardiovascular and respiratory diseases, and the search for new biologically active compounds still remains an urgent problem in pharmacology.

The synthesis of combinatorial libraries on the basis of indolizines had an unexpected continuation. In 2006, Tielmann and Hoenke [8], the chemists of the pharmaceutical company Boehringer Ingelheim, published a detailed research of our discovered recyclization of the oxazolopyridinium salts into aminoindolizines. A vast library of such indolizines was obtained according to Scheme 6.3 and tested for biological activity. In this work, the synthesis technique was optimized (MW oven was used) and many indolizines were found to be unstable. The latter problem was a focus of Tielmann's methodical presentation (the primary concern of combinatorial chemistry is the stability of products) at one of European conferences. We called attention of German researchers to an alternative way to stabilize compounds [9] by introducing an electron-acceptor residue, and this served for making close working contacts between our groups.

Table 6.2 Biological activity of indolizines

Comp. no.	Concentration (μmol)	Binding with receptors, % to control		
		D2-dopamine ^{3}H-QNB	M-muscarine ^{3}H-spiperone	b2-adrenoreceptors ^{3}H-propanolol
XXXIa	100	71.4	152.2	97.0
	10	80.2	96.6	89.7
XXXIb	100	40.7	106.7	84.8
	10	95.6	84.0	75.8
XXXIc	100	100.0	130.9	37.3
	10	95.6	100.4	47.5

6.4   Synthesis of Imidazole Anxiolytics

In the framework of a research of an international INTAS project (no. 00-0711), the PASS program was used for biological activity predictions. Biologists of the Moscow Research Institute of Pharmacology tested potential anxiolytics synthesized by the chemists of Russia, Moldova, Belgium, Greece, and Portugal. The synthetics were allowed to perform computer generation of any combinatorial libraries they were capable to synthesize. By means of the PASS program of 5500 virtual structures, we chose ten, i.e., the most promising ones which were first synthesized and then tested in animal experiments. Mice were exposed to a weak electric shock when attempted to quench thirst. After administration of a potential drug, the fear of shock got weaker, and the frequency of shocks was to be increased.

As a perspective class, we chose condensed imidazoles. The combinatorial library was synthesized as a training task. As seen from Table 6.3, the compounds selected by the program did not only exhibit the predicted activity, but also rank above the standard anxiolitic medazepam. Note that the Moscow group obtained more active compounds than foreign partners, as we mentioned in our joint publication [10]. The imidazoles, previously unknown class, were found to be equally active. This class deserves special mentioning.

6.5   Biologically Active 2-Aminoimidazoles

Shortly after reporting the results on the synthesis of azolyldienes XXIX from azolopyridinium salts (Scheme 6.2a), we focused on the perspective to extend this type of transformation to other fused heterocycles. It was clear that by replacing the pyridine fragment in system **XXXIIa** by the pyrimidine aza analog **XXXIIIb** (cf. reactions a and b in Scheme 6.4), we would obtain, after six-membered ring opening, azadiene XXXIIIb instead of diene **XXXIIIa**.

It was reasonable to expect that C = N bond cleavage in azadiene **XXXIIIb** (Schiff base) would form imidazole **XXXIV** and open-chain fragment.

Table 6.3 Example of the PASS prediction and experimentally observed activity confirmation

Compound no.	PASS prediction (probability)			Activity[a]	Anxiolytic activity of condensed imidazoles
	Activity	To be active	To be inactive		
7.3	Anxiolytic	0.655	0.014	800 ± 128	
	Benzodiazepine ω-agonist	0.670	0.001		
	GABA A receptor agonist	0.424	0.018		

[a]Number of attempts to quench thirst, irrespective of the attendant electric shock

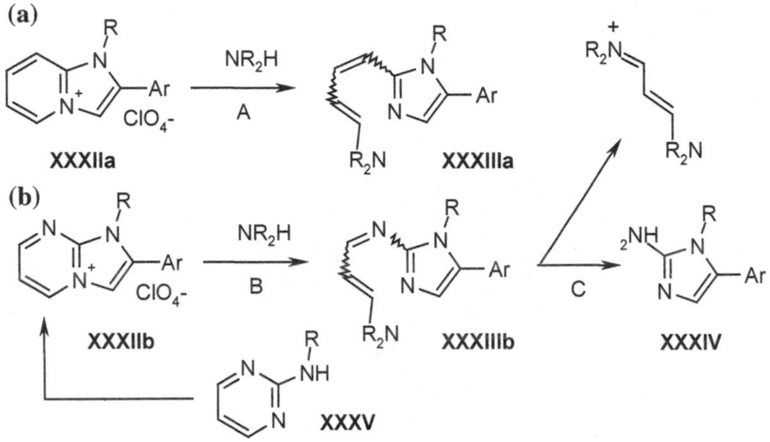

Scheme 6.4 Known **a** and proposed **b** schemes of synthesis of imidazoles from fused azinium salts

2-Aminoimidazoles **XXXIV**, while being structurally simple compounds, had scarcely been reported in the literature. That was roughly the logic of our project proposal in 1999, where we predicted this new reaction just "on the tip on the pen." However, it was not until 2001 that we could, by the trial-and-error method, chose an optimal strategy for the synthesis of imidazoles **XXXIV** from pyrimidines **XXXV** via condensed salts **XXXIIb**: Other strategies (for example, the synthesis of salts **XXXIIb** by alkylation) did not lead to success, and the optimal agent for cleaving the pyrimidine ring proved to be hydrazine. The X-ray diffraction structure for one of the first aminoimidazoles was obtained in 2002 (the graduation thesis of E. Belykh) [11].

We had for long not made public the information on the discovered reaction, waiting for the results of biological tests (in the framework of the INTAS project) and going to patent the scheme of the synthesis and activity of the obtained library. The new patentable methodology formed the subject of open publications, presentations, and dissertations [12]. Probably, the most important application of our reaction, according to our recent findings [13], is the facile full synthesis of natural alkaloids of sea sponges having a 2-aminoimidazole motif in their structure and possessing anti-inflammatory activity. Our protocol involves as little as two–three stages, whereas previously, the full synthesis of such natural substances involved 8–12 (!) stages.

The discovered reactions formed the basis of quite an interesting combinatorial project. According to the PASS prediction, aminoimidazoles X might exhibit antiprotozoal activity and be effective against leishmaniasis (a tropical fever), an extremely heavy disease in Third World countries. The search for biological laboratories focused on tropical fevers led us to discovery and further fruitful cooperation with the University of Karachi (Pakistan). The first results showed that our imidazoles are comparable in the antileishmaniasis activity with the standard drug amphotericin (the disadvantage of the latter is its high toxicity). Searching for further funding for this project, we submitted a Russian–Pakistanian bilateral project proposal to the World Health Organization (WHO). Even though the WHO uses to fund early stages of search for drugs against tropical diseases, we were recommended to accumulate and extend evidence on the biological and ADME properties of our compounds. As a result, the whole project acquired a clear goal and quite obvious ways to reach this goal.

Note that the synthesis of imidazole libraries by Scheme 6.4 is fairly easy, since its all three stages form readily purified precipitates. In this connection, the reaction was adapted for the SynCore apparatus as a training task for the special laboratory course in combinatorial chemistry at the MSU. Students and teachers prepared there a large library of imidazoles. In the framework of the joint educational project with the Chemical Diversity Research Institute (CDRI, Khimki), we performed a demonstrative robotized screening of a series of imidazoles. In particular, we measured their primary properties of key importance for potential drugs (cytotox icity, membrane transport ability, solubility, etc.).

Later, the MSU concluded contracts first with a small pharmaceutical company Pacific Pharma Technologies and then with a larger venture company Upstream Bioscience on the synthesis of big libraries of imidazoles and their comprehensive screening. Initially, our compatriot Prof. A. Cherkasov performed in vitro experiments at specialized laboratories in Canada. Later, the African physician Prof. G. Olobo (University of Campala, Uganda) initiated in vivo experiments in "field" conditions. Even though the global financial crisis has made problematic finalization of the experiments planned after 2010, the available evidence instills cautious optimism.

6.6 Other Examples of Practical Application of New Reactions

The four examples considered above illustrate an adaptation to applied tasks of a purely academic approach (design of previously unknown reactions and new families of compounds) to the design of useful properties. Therewith, the PASS program fulfilled its heuristic role. Now, we would like to mention three more examples from our practice briefly, in which the structural novelty of compound classes (or transformations) served as a basis for quite pragmatic applied projects directly related to combinatorial chemistry.

1. **New Color Test for Amines**. By reacting azoloazinium salts with amines (Scheme 6.4), we noted an intense violet color of dienes in those cases where the five-membered ring contained a paranitrophenyl residue. Apparently, in the latter cases, quite an extended conjugation chain arises: from the amino to nitro group via diene, azole, and phenyl ring. Still more intense (almost black) coloration under the action of amines was observed with imidazopyrimidinium salt

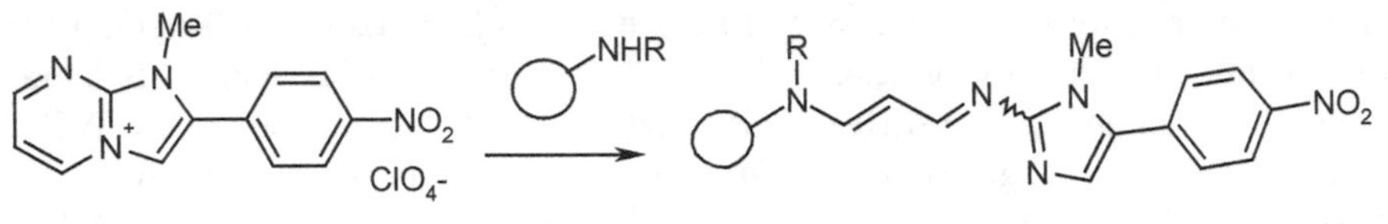

Scheme 6.5 Patented color reaction of fused imidazopyrimidinium salts with amines

with the same para-nitrophenyl group, due to the formation of analogous aza-diene (Scheme 6.5).

Practically, a new highly sensitive color test for secondary amines was discovered, which was superior by some parameters than the standard ninhydrin test (Kaiser test). This test was found to be the most valuable for solid-phase synthesis, since it allowed free amino groups to be determined on the resin, even in the presence of thiols. We have patented this result together with researchers of the Catholic University of Leuven [14].

2. **"Inorganic" Bioactivity Component**. The second example illustrates the fact that an unexpected factor may affect biological activity in a series of one-type compounds. We found in our early work [15] that the reaction of 2-chloropyridinium salts **XXX** with potassium thiocyanate proceeds with a smooth closure of the thiazole ring to form condensed ionic systems **XXXVI** (Scheme 6.6). Usually, all the three atoms of thiocyanates (S–C–N) are involved in thiazole ring formation. In our discovered reaction, the thiocyanate nitrogen was not incorporated into the new thiazole ring. Hence, the structural novelty of the reaction is clear.

Salts **XXXVI** showed activity in certain agrochemical tests. This activity varied incomprehensibly and quite dramatically: from necrosis of seeds to the enhancement of their germinating ability. In our bilateral project with the Japanese agrochemical concern Nippon Soda, we undertook a more detailed study of structure–activity correlations. It was found [16] that the nature of the aryl group (the only varied residue) is almost insignificant. The key role here, incredible as it may seem, belongs to the nature of the inorganic counterion in the obtained salts. To find out what proportion between chloride, bromide, and thiocyanate in salts **XXXVI** affects bioactivity, we applied anionic chromatography, a technique typical of analysis of mineral substances.

3. **Principle of "Supereconomic" Reagents in Combinatorics**. The third example relates to new perspectives of practical use of structurally unusual compounds: mesoionic compounds. We have for long studied the chemistry of the entire class of mesoionic systems under financial support of the Russian Foundation for Basic Research. During the research into the chemistry of mesoionic oxazoles **XXXVII** (bicyclic munchnones), we noted their possibility of simultaneous recyclization into salts **XXXVIII** (Scheme 6.7) [1, 17]. It had long been not quite obvious for us what benefits can be gained by

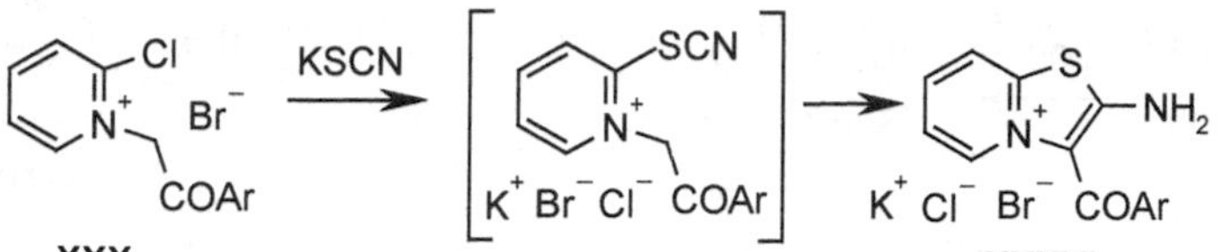

Scheme 6.6 Discovered thiazole ring closure reaction

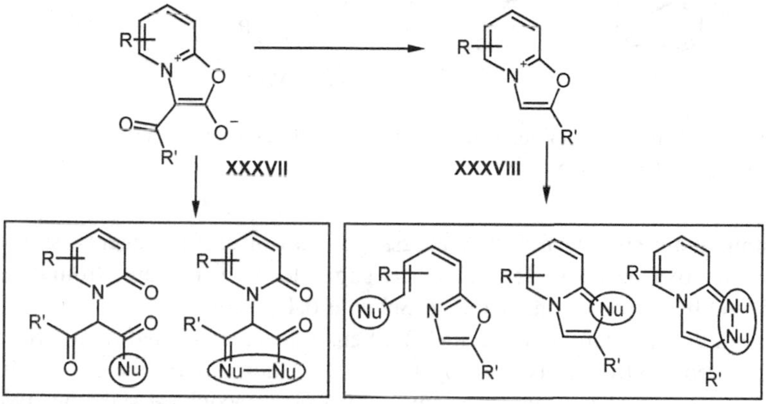

Scheme 6.7 Schematic representation of ring transformations mesoionic oxazoles (*left*) and derived from the fused oxazolium salts (*right*)

pharmaceutical industry from the transformation of one labile system into another (**XXXVII** into **XXXVIII** in Scheme 6.7).

This had been the case until Bayer announced its memorable Synthon project [18]. The goal of this project was to create the world's largest bank of rare reagents, so-called synthons. We turned attention of Bayer's chemists to the fact that the transformation in Scheme 6.7 is the most economic, since it represents a rare example of this a synthon-to-synthon transformation. Actually, munchnones **XXXVII** can be reacted with various nucleophiles (and binucleophiles) to obtain large collections of diverse compounds **XIII** (Scheme 6.7). If these synthons and products are no longer interesting, old synthons (**XXXVII**) can be easily converted into new ones (**XXXVIII**) and obtain new-type libraries, using the same nucleophiles. The "Synthon-from-Synthon" project formed the basis of long-year cooperation between the MSU and Bayer and was continued until radical reorganization of the company.

As seen, certain synthetic ideas which, at first glance, seemed to be quite far from the practice of combinatorial chemistry could be materialized in concrete libraries of biologically active compounds. Concluding the brief review of these projects, we would like to mention some other experimental findings (Scheme 6.8) which are direct candidates of combinatorial chemistry applications. First, this is an

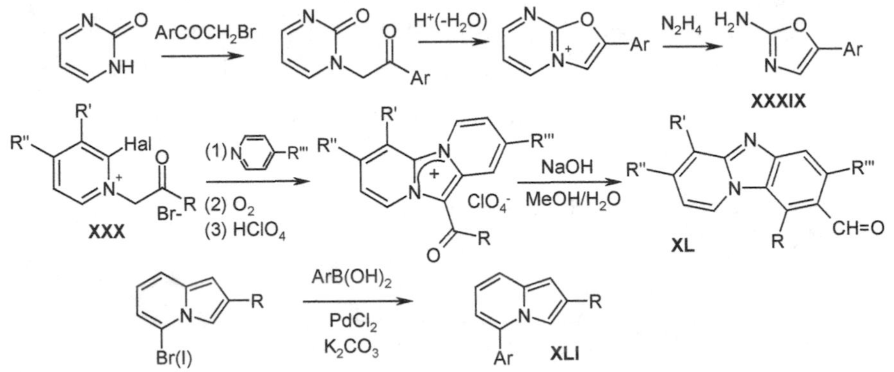

Scheme 6.8 Different experimental findings in the area of heterocyclic synthesis applicable in the libraries design of biologically active compounds

interesting and a previously hardly accessible class of useful reagents, substituted with 2-aminooxazoles **XXXIX** which were suggested to be prepared from the cheap pyrimidone by a scheme analogous to Scheme 6.4 [19].

Second, this is a family of pyrido[1,2-a]benzimidazole aldehydes **XL** which can be easily prepared from salts **XXX** by successive action of pyridine and alkali [20]. This tricyclic system is a typical example of a pharmacophore skeleton: Even its simplest parent structure is analgesic, and no universal convenient synthetic approach to functional derivatives of such scaffold is still known. The third family is the parallel liquid-phase synthesis of the library of 5-arylindolizines **XLI** [21] and aza analogs by the cross-coupling reaction. These indolizines are quite an interesting type of indicators: They exhibit a strong fluorescence with a high quantum yield, and, therewith, change the radiation color on protonation (due to a radical rearrangement of their π system).

6.7 Conclusions

Many pharmaceutical companies have to cut down their expenses for combinatorial projects (may be due to financial reasons), which resulted in a fall of the general interest in this field among synthetic chemists. It is noteworthy that the domestic pharmaceutical industry is developing in a specific way, and combinatorial chemistry in Russia is not excluded to experience the blossoming period in the future. As positive long-standing trends, we can mention, for example, the Program of Revival of Domestic Pharmaceutical Industry until 2020 (Pharma-2020 project), which includes, as an essential constituent part at the early stage, the development of projects in combinatorial chemistry. Other positive tendencies include the growing interest of State Corporations in the problems of search for new biologically active compounds and its associated infrastructure. The main guarantee for successful future projects in combinatorial chemistry is, in our opinion, qualified research staff which have not only been lost over the past decade, but also, by contrast, gained quite a qualified training.

Acknowledgements The author is grateful to Dr. D. Ermolat'ev (Leuven, Belgium), Prof. V. Scott (USA, Indiana), and Dr. Yu. Sandulenko (Fistech, Moscow) for consultations and procedures in SPOS methodology. I am also thankful to Drs. M. Mironov (Ekaterinburg) and N. Ivanova (Chembridge) for LPOS methodology. The author thanks all the members of his research group at MSU (Drs. I. Dlinnykh, V. Alifanov, A. Bush, A. Tsisevich, A. Efimov, D. Al'bov, O. Mazina, Mrs. E. Belykh, and A. Nevskaya) and Dr. V. Rybakov for their help.

References

1. (a) Babaev EV (2007) Doctoral Habil (Chem) Dissertation, Moscow. (b) Babaev EV (1997) In: Attanasi OA, Spinelli D (eds) Targets in heterocyclic systems—chemistry and properties. Soc Chim Ital, Rome. (c) Babaev EV, Zefirov NS (1996) Strategy and tactics of computer-assisted forecast of novel recyclizations in the series of azolopyridines with bridgehead nitrogen atom. Chem Heteroc Comp 32(11–12): 1344-1357. doi:10.1007/BF01169965. (d) Babaev EV (2000) Fused munchnones in recyclization tandems. J Heterocycl Chem 37: 519–526. doi:10.1002/jhet.5570370309. (e) Babaev EV, Alifanov VL, Efimov AV (2008) Oxazolo[3,2-a]pyridinium and Oxazolo[3,2-a]pyrimidinium Salts in Organic Synthesis. Russ Chem Bull (Intern Ed) 57(4): 845–862. doi:10.1007/s11172-008-0122-8
2. (a) Filimonov DA, Poroikov VV (1996) Bioactive compounds design: possibilities for industrial use. BIOS Sci Publ, Oxford. (b) Poroikov VV, Filimonov DA, Ihlenfeld WD, et al. (2003) PASS biological activity spectrum predictions in the enhanced open NCI database browser. J Chem Inf Comput Sci 43: 228–236. doi:10.1021/ci020048r. (c) www.way2drug.com
3. (a) Babaev EV, Efimov AV, Maiboroda DA, Jug K (1998) Unusual ambident behavior and novel ring transformation of Oxazolo[3,2-a]pyridinium salts. Eur J Org Chem 1: 193–196. doi:10.1002/(SICI)1099-0690(199801)1998:1<193::AID-EJOC193>3.0.CO;2-6.
(b) Maiboroda DA, Babaev EV, Goncharenko LV (1998) 1-Amino-4-(5-arylazol-2-yl)-1,3-butadienes: synthesis and study of spectral and pharmacological properties. Pharm Chem J 32(6): 310–314. doi:10.1007/BF02580516
4. Babaev EV, Tsisevich AA (1999) Unusual ring transformation of 2-Halogen-N-Phenacylpyridinium salts in reaction with secondary amines: a facile "One-Pot" route to 1-Amino-4-(Oxazolyl-2)-Butadienes. J Chem Soc Perkin Trans 1(4):399–401. doi:10.1039/a809694e
5. Babaev EV, Nasonov AF (2001) Formation of oxazoles from 2-methylsulfanyl-N-phenacylpyridinium salts. ARKIVOC 2: 139–145. www.arkat-usa.org/get-file/19188/
6. Babaev EV, Efimov AV (1997) Novel approach to synthesis of indolizine nucleus ad the class of 5-aminoindolizines via recyclization of 5-methyloxazolo[3,2-a]pyridinium cation: the next confirmation of computer forecast. Chem Heteroc Comp 33(7):875–876. doi:10.1007/BF02253046
7. (a) Babaev EV, Efimov AV, Tsisevich AA, Nevskaya AA, Rybakov VB (2007) Synthesis of 5-alkoxyindolizines from oxazolo[3,2-a]pyridinium salts. Mendeleev Commun 17(2): 130–132. doi:10.1016/j.mencom.2007.03.027. (b) Babaev EV, Tsisevich AA, Al'bov DV, Rybakov VB, Aslanov LA (2005) Heterocycles with a bridgehead nitrogen atom. Part 16. Assembly of a peri-fused system from an angular tricycle by recyclization of an oxazole ring into pyrrole one. Russ Chem Bull (Int Ed) 54(1): 259–261. doi:10.1007/s11172-005-0248-x.

(c) Efimov AV (2006) Development of synthetic methods leading to 5-substituted indolizines. Cand Sci (Chem). Dissertation, Moscow University. (d) Tsisevich AA (2006) Novel ring transformations of mono-, bi- and tricyclic analogs of oxazolo[3,2-a]pyridines. Cand Sci (Chem) Dissertation, Moscow University. (e) Al'bov DA (2005) X-ray structures of cycloalcano[b]pyridones and their heterocyclization products. Cand Sci (Chem) Dissertation, Moscow University. (f) Mazina OS (2005) Crystal structure of novel condensed tricyclic systems from nitriles of cycloalkanopyridine. Cand Sci (Chem) Dissertation, Moscow University

8. Tielmann P, Hoenke C (2006) Optimization, scope and limitations of the synthesis of 5-aminoindolizines from oxazolo[3,2-a]pyridinium salts. Tetrahedron Lett 47:261–265. doi:10.1016/j.tetlet.2005.11.033

9. (a) Mazina OS, Rybakov VB, Chernyshev VV, Babaev EV, Aslanov LA (2004) X-ray Mapping in heterocyclic design: XII. X-ray diffraction study of 2-pyridones containing cycloalkane fragments annelated to the C(5)–C(6) bond. Crystallogr Rep 49(2):158–168. doi:10.1134/1.1756641. (b) Babaev EV, Vasilevich NI, Ivushkina AS (2005) Efficient synthesis of 5-substituted 2-aryl-6-cyanoindolizines via nucleophilic substitution reactions. Beilstein J Org Chem 1: 9. http://bjoc.beilstein-journals.org/content/pdf/1860-5397-1-9.pdf. (c) Rybakov VB, Babaev EV (2014) Transformations of substituted Oxazolo[3,2-a]Pyridines to 5,6-disubstituted indolizines: synthesis and X-ray structural mapping. Chem Heterocyc Comp 50(2):225–236. doi:10.1007/s10593-014-1465-8

10. Geronikaki A, Babaev E, Dearden J, Dehaen W, Filimonov D, Galaeva I, Krajneva V, Lagunin A, Macaev F, Molodavkin G, Poroikov V, Pogrebnoi S, Saloutin V, Stepanchikova A, Stingaci E, Tkach N, Vlad L, Voronina T (2004) Design, synthesis, computational and biological evaluation of new anxiolytics. Bioorg Med Chem 12:6559–6568. doi:10.1016/j.bmc.2004.09.016

11. Rybakov VB, Babaev EV, Belykh EN (2002) 2-Amino-5-(4-chlorophenyl)-1-methylimidazole. Acta Crystallogr Sect E 58:o126–o128. doi:10.1107/S1600536802000284

12. (a) Ermolat'ev DS, Babaev EV, van der Eycken EV (2006) Efficient one-pot, two-step, microwave-assisted procedure for the synthesis of polysubstituted 2-aminoimidazoles. Org Lett 8(25): 5781–5784. doi:10.1021/ol062421c. (b) Ermolat'ev DS, Svidritsky EP, Babaev EV, van der Eycken E (2009) Microwave-assisted synthesis of substituted 2-amino-1H-imidazoles from imidazo[1,2-a]pyrimidines. Tetrahedron Lett 50(37):5218–5220. doi:10.1016/j.tetlet.2009.06.128. (c) Ermolat'yev D (2008) Doctoral (Chem) Dissertation, Leuven

13. Ermolat'ev DS, Alifanov VL, Rybakov VB, Babaev EV, van der Eycken EV (2008) A concise microwave-assisted synthesis of 2-aminoimidazole marine sponge alkaloids of the isonaamines series. Synthesis 13:2083–2088. doi:10.1055/s-2008-1078444

14. Patent WO2008EP58537

15. Babaev EV, Bush AA, Orlova IA, Rybakov VB, Zhukov SG (1999) Novel route to b-fused thiazoles starting from a 2-chloro-1-phenacylpyridinium salt and KSCN. Crystal structures of thiazolo- and oxazolo[3,2-a]pyridinium thiocyanates. Tetrahedron Lett 40(42):7553–7556. doi:10.1016/S0040-4039(99)01601-9

16. Babaev EV, Bush AA, Orlova IA, Rybakov VB, Ivataki I (2005) New mesoionic systems of azolopyridine series. Part 2. Synthesis, structures, and biological activity of 2-aminothiazolo [3,2-a]pyridinium salts and thiazolo[3,2-a]pyridinium 2-imidates. Russ Chem Bull (Int Ed) 54 (1):231–237. doi:10.1007/s11172-005-0242-3

17. (a) Babaev EV, Orlova IA (1997) Novel synthesis of oxazolo[3,2-a]pyridinium cation by recyclization of its mesoionic precursor: confirmation of computer forecast. Chem Heteroc Comp 33(4):489. doi:10.1007/BF02321398. (b) Kazhkenov ZM, Bush AA, Babaev EV (2005) Dakin-west trick in the design of novel 2-alkyl(aralkyl) derivatives of oxazolo[3,2-a] pyridines. Molecules 10(9):1109–1118. doi:10.3390/10091109. (c) Rybakov VB, Babaev EV, Chernyshev VV (2002) X-ray mapping in heterocyclic design: VII. Diffraction study of the structure of N-pyridoneacetic acid and the product of its intramolecular dehydration. Crystallogr Rep 47(3):428-432. doi:10.1134/1.1481930

18. Babaev EV (2000) Abstracts of papers, workshop on organic chemistry and catalysis (RAS–Bayer), Moscow, 7–8 Feb 2000
19. Alifanov VL, Babaev EV (2007) Novel and efficient synthesis of 2-aminooxazoles from pyrimidine-2-one. Synthesis 2:263–270. doi:10.1055/s-2006-958941
20. Babaev EV, Tikhomirov GA (2005) Heterocycles with a bridging nitrogen atom, Part 15. The novel recyclization of dipyrido[1,2-a:1′,2′-c]imidazolium salts to pyrido[1,2-a]benzimidazole-8-carboxaldehydes. Chem Heteroc Comp 41(1):119–123. doi:10.1007/s10593-005-0117-4
21. (a) Kuznetsov AG, Bush AA, Rybakov VB, Babaev EV (2005) An improved synthesis of some 5-substituted indolizines using regiospecific lithiation. Molecules 10(9):1074–1083. doi:10.3390/10091074. (b) Kuznetsov AG, Bush AA, Babaev EV (2007) Synthesis and reactivity of 5-Br(I)-Indolizines and their parallel cross-coupling reactions. Tetrahedron 64(4): 749–756. doi:10.1016/j.tet.2007.11.017. (c) Shadrin IA, Rzhevskii SA, Rybakov VB, Babaev EV (2015) Sonogashira reaction of the indolizine ring. Synthesis 47:2961–2964. doi: 10.1055/s-0034-1378861

Printed in the United States
By Bookmasters